# Science
## from the beginning

# Science
## from the beginning

B L Hampson and K C Evans

*Teacher's Book 2*

OLIVER & BOYD

**OLIVER and BOYD**
Robert Stevenson House, 1–3 Baxter's Place
Leith Walk, Edinburgh EH1 3BB

*A Division of the Longman Group Limited*

FIRST PUBLISHED 1961
NEW EDITION 1977
THIRD IMPRESSION 1983

ISBN 0 05 002928 2

Printed in Hong Kong by Wah Cheong Printing Press Ltd.

4

# Contents

Lesson                                                           Page

Introduction                                                  7
     General Scheme of Work                  12
     Keeping Specimens for Observation Purposes   20
     Useful Materials to Have in Stock       30
     The Use of the Science Table           34
1   The Three Forms of Never-Alive Things     38
     Solid, Liquid and Gas
2   The Four Main Needs of Living Things      44
     Oxygen: Food: To Grow: To Have Young
3   Living Plants on Land                  51
     Herbs: Trees: Shrubs: Simple Plants
4   Parts of Herbs, Trees and Shrubs        57
     Flowers, Fruits and Seeds
5   Animals With Six Legs                68
     Living Insects Grow Up
6   Take Care of Living Things           81
     Let Insects Have Oxygen and Food
7   Living Animals Can Move From Place to Place   85
     Moving in Water, on Land and in the Air
8   Other Things Are Moved from Place to Place   92
     By Moving Animals, Moving Gases and Moving Liquids
9   Forcing Things to Move              96
     How We Move Things from Place to Place on Land
10 Moving Things from Place to Place on Land   103
     Rollers and Wheels Help
11 The Three Weather Makers            108
     Sun, Air and Water
12 Forcing Things to Move              114
     The Earth Pulls
Questions on Lessons 1 to 12           122
13 Solids, Liquids and Gases            125
     Mixtures of Things
14 Mixing Things                    131
     Different Kinds of Mixtures
15 A Liquid We Need                 137
     Water

16  Water in Winter                                              145
      Freezing and Melting
17  A Solid We Need                                              150
      Common Salt
18  A Gas We Need                                                155
      Oxygen Gas
19  Forcing Things To Move                                       163
      Magnets Pull
20  Herbs, Trees and Shrubs Grow Bigger                          170
      Stems and Buds
21  Living Things Have Young                                     175
      The Animals
22  Some Water-and-Land Animals                                  183
      Amphibians
23  Animals With Scaly Skins and Lungs                           198
      Reptiles
24  Living Things Have Young                                     211
      The Plants
Questions on Lessons 13 to 24                                    219
25  Herbs, Trees and Shrubs With Flowers                         221
      Flowers on Their Own and Flowers in Clusters
26  Leaves of Herbs, Trees and Shrubs                            228
      Leaves and Leaflets
27  Stems and Roots of Herbs, Trees and Shrubs                   234
      Above the Ground and Below
28  Insects With a Four-Stage Life                               243
      Moth and Butterflies
29  Insects With a Four-Stage Life                               251
      The Two-Winged Flies
30  Never-Alive Things in Space                                  259
      Stars, Planets and Satellites
31  Dead Things Are Used By Living Things                        273
      The Main Uses For Dead Things
32  Rock, Wind and Water                                         279
      Rock Is Worn Down
33  Living Plants in Fresh Water                                 286
      Herbs and Simple Plants
34  Animals With Eight Legs                                      295
      Spiders and Harvestmen
35  Animals With Crusty Skins                                    305
      Crustaceans
Questions on Lessons 25 to 35                                    314
36  Land Animals With Many Legs                                  316
      Centipedes and Millepedes
Index                                                            324

# INTRODUCTION

Science is a systematic way of examining things and occurrences. Very simply, it is a study of:

1 the different kinds of things which exist,
2 what happens to them.

The aim of this series of books for the junior school is to provide:

1 a logical beginning for a general science training,
2 the establishment, by progressive stages, of a comprehensive foundation of general scientific knowledge from which any secondary phase of science education may be developed.

# CONTENTS OF TEACHER'S BOOK 2

This consists of thirty-six lessons, each one corresponding to a double-page section in Pupils' Book 2. Each of these lessons is divided into five parts:

1 Demonstration material
2 Sample link questions
3 Relevant information
4 CODE (Collection, Observation, Demonstration, Experiment)
5 Written work

**Demonstration Material**

Under this heading will be found suggestions for apparatus or specimens suitable for illustrating the particular lesson.

*Apparatus*

Where this is listed, it is in most cases simple and of a kind familiar to children in their own homes, *e.g.* jam jar, candle, tin lid.

*Specimens*

*Alive*. The value of keeping living specimens in the classroom is

7

obvious. Where these could usefully illustrate a lesson, suggestions are given for suitable ones, and also information on maintenance.

*Dead.* For many lessons dead specimens are useful. Some animal and plant parts – for example, feathers and wood – do not deteriorate rapidly and therefore can be easily kept. Others – for example, a tadpole – decompose quite quickly, and preserved specimens of these have been found invaluable. They are of even more interest when the children themselves have helped with the preserving. Included in this Introduction is a section dealing with simple methods of preserving animal and plant material, which may thus be kept indefinitely and used year after year, if required.

*Never Alive.* Various solids and liquids are not only useful as material for illustrating many lessons, but, together with preserved dead specimens, form a useful nucleus for any class or school science museum. It is better, of course, if they are simply, but meaningfully, labelled.

### Sample Link Questions

Although most teachers will undoubtedly apply their own methods of establishing lesson-continuity, based on their personal knowledge of the individual class, certain revisionary questions are included at the beginning of each lesson. These provide a link with previous lessons on the same subject, or on interrelated subjects, and are based on questions which have been found useful in actual practice. They are not exhaustive but merely serve as a basis throughout the book for a systematic and constant revision of the most important points outlined during the successive stages which the course pursues. When the answer to a question has been discovered during a previous lesson, it is printed, for convenience, beside the question.

In addition to these sample link questions, three lists of questions have been included at certain stages through the book. Between them they cover the most important points and are phrased in such a way that in many cases they require an answer of only one or two words. This permits, where desired, of their being answered in writing.

## Relevant Information

This is intended to serve two purposes:

1 to act as a source of material for that particular lesson,
2 to provide information and facts useful to the teacher as background knowledge.

## CODE

Under this heading will be found suggestions for:

*Collection* of material which would suitably illustrate that particular lesson. On page 20, under the title *Keeping Specimens for Observational Purposes*, will be found information on suitable methods of retaining material for indefinite use.

*Observations* which children may be encouraged to make – not necessarily at school, *e.g.* observing how the wheels are attached to toy motor cars and other toys. Observation is the start of many a scientific endeavour.

*Demonstrations* to emphasise a particular point. This is what is sometimes loosely termed 'experimenting'. There are two main ways in which this may be undertaken, according to the discretion of the teacher:

1 by the teacher alone, where the demonstration material is such that it may be inexpedient for the children to handle it,
2 by the teacher with children assisting.

*Experiments* which may be undertaken by the class as a whole, or by children working individually or in groups. It is recommended that from the start the principle of controlled experiment should be emphasised, if the enquiring mind of a child is to be trained to pursue its enquiries scientifically. For example, if a damp cloth is placed on a radiator to show that heat helps water to become water vapour, then a similar damp cloth of the same size should be placed nearby, so that a comparison can be made.

## Written Work

This is in effect the 'Answer section', and assists in easy marking of the written work set in the Pupils' Book.

9

# PUPILS' BOOK 2

This book consists of thirty-six double-page lessons. It follows the assumption that even to young children, science will be an affair of observing and finding out first, and reading for confirmation last. Each section is divided into five parts:

1 A page of illustrations in full colour.
2 A simple sentence summary of the main points of the lesson.
3 A part entitled 'For you to draw' in which suggestions are given for notebook drawings. Sometimes it is best for children to draw an actual specimen which is in front of them, *i.e.* from their own observations. In addition, there are obvious advantages in children being encouraged to make their own 'field drawings' of things which they experience outside the school.
4 A part entitled 'For you to write' in which simple puzzle sentences are provided. These are sentences from which key words have been omitted, and which need to be thought out by the individual pupil, using the text as a guide.
5 A part entitled 'For you to do', in which are suggestions on what the class can collect, observe, and experiment with, relative to that particular lesson and subsequent lessons.

The number of the lesson is shown on a blue spot at the top of the left-hand page.

## The Scheme

The scheme for each year is divided into three sections:
1 The science table
2 Classification section
3 General subjects

## The Science Table

It is with this that the introduction to scientific study begins. It begins, not so much by assembling a haphazard collection of items and attributing to them names, but by classifying those items from the start according to whether they are *alive, dead,* or *never alive.* Thus the question which a child needs to learn to ask first, when encountering something entirely new to his or her experience, is not

10

'What is it called?', but 'What kind of a thing is this?'. A separate section on the use of the science table will be found on pages 34–37.

By this method of classifying objects into groups first, and naming the known individuals last, the science table also provides an introduction to the classifications which follow.

## Classification Section

The narrower interpretations of nature study may stress the study of certain selected objects – say, a dandelion or a frog. Learning numbers of facts about these particular things does not, however, enable a child to acquire any general knowledge of plants or of amphibians.

The answer to the question 'What is it called?' can only be supplied by someone who happens to have learnt what it is called. The question 'What kind of thing is it?' can be answered simply by a child mind equipped with a knowledge of the characteristics of groups of things. This is not intended to imply that a junior should be expected to examine with an analytical mind from the start. But he can be encouraged to examine in a logical way, and be systematically provided with the knowledge of how to do so. For example: a child encounters a living object. It is observed to move from one place to another, therefore it is established as a living animal, and not a living plant. The next observation is that it moves on six legs, therefore it is further established as an insect. One of the aims of this scheme is to teach the observable characteristics of groups of things, so that any plant, any animal, or any never-alive thing may be examined with reasoning.

The number of classifications to be learnt increases with successive years, so that during each year, established classifications may be consolidated and further sub-classifications introduced.

## General Subjects

Lessons under this heading are of three main kinds:

1 Lessons to illustrate classifications. These are based on accepted generalisations.
2 Lessons on interesting topics.
3 Lessons on observable phenomena or 'happenings'.

The fundamental aim of any living species is the propagation of its own kind. It will be found that where topic lessons are about individual living organisms, they generally centre round their four basic needs – oxygen, food, to grow, to have young.

# GENERAL SCHEME OF WORK IN BOOK 2

| Subject | Topic | Collection of Material by Children, and Local Observation | Demonstration and Experiment |
|---|---|---|---|
| (A) *The Science Table* Collected items to be separated into seven distinctly named groups | 1 Alive<br>  *a* Animal<br><br><br><br>  *b* Plant | *E.g.* Fish, amphibians, reptiles, insects, spiders etc in apropriate containers<br>*E.g.* Specimens from herbs, trees and shrubs, such as roots, stems, leaves, flowers, fruits and seeds. Specimens of simple plants such as algae, fungi, mosses | |
| | 2 Dead<br>  *a* Animal<br><br><br><br><br>  *b* Plant | *E.g.* Mounted or preserved specimens, together with dead animal parts such as fur, feathers, leather, reptile skin, bones<br>*E.g.* Mounted or preserved specimens, together with dead plant parts, such as wood, cork, bark, peat, autumn leaves, various plant fibres | |
| | 3 Never alive<br>  *a* Solids<br><br><br><br><br><br><br>  *b* Liquids<br><br><br><br>  *c* Gases | *E.g.* Various rocks and ores, salt, metals, glass, plastics, and various manufactured articles containing never-alive substances, mollusc shells<br>*E.g.* Water, ink, oil, turpentine, paraffin, methylated spirit, mercury,. mixtures in liquid form<br>*E.g.* 'Empty' jars with lingering smells like those from perfume, pickles etc | |

| Subject | Topic | Collection of Material by Children, and Local Observation | Demonstration and Experiment |
|---|---|---|---|
| **(B) *Classification Section*** | | | |
| 1 The three forms of never-alive things | *a* Solid<br>*b* Liquid<br>*c* Gas | Collect various solids and liquids. Observe fixed shape and size of solids | Demonstrate: *a* that a liquid has no particular shape; *b* that a gas has neither shape nor size |
| 2 Plants on land | *a* With roots, stems and leaves<br>  *1* Herbs<br>  *2* Trees<br><br><br>  *3* Shrubs<br><br><br>*b* Without roots, stems and leaves *i.e.* simple plants | <br><br>Stems not woody<br>One main woody stem growing from the roots<br>More than one main woody stem growing from the roots<br>Examples of algae, mosses and fungi | |
| 3 Plants in fresh water | *a* With roots, stems and leaves<br>Herbs only<br><br><br><br><br><br><br><br><br>*b* Without roots, stems and leaves, *i.e.* simple plants | *1* Swamp herbs, *e.g.* rushes, reedmace<br>*2* Herbs with floating leaves, *e.g.* water lily, water buttercup<br>*3* Floating herbs, *e.g.* frogbit, duckweed<br>*4* Underwater herbs, *e.g.* Canadian waterweed, hornwort<br>*1* Algae, *e.g.* pond slime, blanket weed<br>*2* Great water moss<br>*3* Fungi, *e.g.* fungus on fish | |
| 4 Additional parts found only on herbs, trees and shrubs | *a* Flowers<br>*b* Fruits<br>*c* Seeds | Collect flowers<br>Collect autumn fruits and seeds. Observe that a true fruit completely encloses the seed | Cut open a fruit to expose the seed or seeds inside |
| 5 Stems and buds | *a* End buds and side buds<br>*b* What grows from buds<br>  *1* branch stems with new leaves<br>  *2* flowers<br>  *3* branch stems bearing both leaves and flowers | *a* Observe what grows from buds<br>*b* Observe position of new bud in axils of leaves | *a* Grow twigs in water, bruising ends first<br>*b* Examine smaller buds inside a sprout or cabbage bud |

| Subject | Topic | Collection of Material by Children, and Local Observation | Demonstration and Experiment |
|---|---|---|---|
| 6 Flowers | a On their own | E.g. daffodil, wood anemone, deadly nightshade | |
| | b In clusters | E.g. catkins, laburnum, bluebell, grass flowers, dandelion | |
| 7 Different kinds of leaves | a Simple (one blade) 1 without a stalk | E.g. scarlet pimpernel, canterbury bell | Grow twigs standing in water |
| | 2 with a stalk | E.g. beech, lime, rhubarb etc | |
| | b Compound (more than one blade) 1 with leaflets at end of stalk | E.g. lupin, horse chestnut, laburnum | |
| | 2 with leaflets at side of stalk | E.g. ash, elder, fern | |
| 8 Plants have young | a By part of the parent growing into a new plant | E.g. strawberry, twigs with adventitious roots | Grow potato on jar of water |
| | b By spores | E.g. mushroom and toadstool, mosses, ferns | Obtain a 'spore print' from a toadstool |
| | c By seeds | E.g. cones and seeds, and fruits and seeds | a Cut open a bean seed to expose plant inside b Grow herbs from seeds |
| 9 Animals have young | a By part of the parent growing into a new animal | Skeleton remains of sea mat, coral, sponge | |
| | b By means of eggs | Various eggs and egg cases | |
| | c By means of young not contained in an egg | Baby mammals such as hamsters, kittens, mice, puppies | |
| 10 Amphibians | a Descent from fish | British amphibians lay eggs in water as fish do | a Keep eggs and tadpoles in aquaria |
| | b Smooth moist skin, or rough dry skin | Adult frogs, toads and newts | b Keep adults in vivaria |
| | c Life in water and on land | Egg, tadpole, adult | |
| | d Tailed, tailless, and legless | Adults | |
| 11 Reptiles | a Descent from amphibians | a Observe scales on skin of any reptile | |
| | b Scaly skins and lungs | b Compare with smooth moist skin or rough dry skin of an amphibian | |

| Subject | Topic | Collection of Material by Children, and Local Observation | Demonstration and Experiment |
|---|---|---|---|
| 11 Reptiles (cont.) | c Eggs laid on land<br>d Pre-historic monsters<br>e Modern reptiles<br>  1 crocodiles<br>  2 turtles, terrapins, tortoises<br>  3 lizards<br>  4 snakes | | |
| 12 Life histories of insects | a Some have a three-stage life<br><br>b Most have a four-stage life | Observe larvae similar to adults, *e.g.* earwig, grasshopper, dragonfly, stick insect<br>Observe larvae unlike the adults, *e.g.* moth, butterfly, two-winged flies, beetles | Retain stick insect eggs for later hatching<br><br>Rear bluebottle larvae to adult stage |
| 13 Moths and butterflies | a Adults covered with powdery scales<br>b Four-stage life<br><br>c 1 Adult butterflies have clubbed antennae and rest with wings up<br>  2 Adult moths rest with wings down | Any adult moth or butterfly<br>Egg, caterpillar, chrysalis, adult<br>Observe antennae of adults<br>Observe position of wings on adults at rest | Rear larvae in suitable containers |
| 14 Two-winged flies | a Front wings for flying<br>b Rear wings reduced to balancers<br>c Four-stage life<br>d Carrying diseases | Adult crane flies, houseflies, bluebottles | Rear larvae *e.g.* gnat, bluebottle, in suitable containers |
| 15 Spiders and harvestmen | a Spiders<br>  1 Eight legs<br>  2 Body divided by a waist<br>  3 Hunters or trappers<br>  4 Silk for cocoons and webs<br>b Harvestmen<br>  1 Eight legs<br>  2 No waist<br>  3 Hunters<br>  4 No silk | Observe<br>a eight legs<br>b waist<br>c no wings<br>d use of silk<br><br><br>Observe<br>a eight legs<br>b no waist<br>c no wings<br>d no silk | Keep in suitable containers |
| 16 Crustaceans | a Crusty skins<br>b Four feelers<br>c Some have claws | In sea water: crab, lobster, prawn, shrimp | Keep in suitable containers |

| Subject | Topic | Collection of Material by Children, and Local Observation | Demonstration and Experiment |
|---|---|---|---|
| 16 Crustaceans (cont.) | | In fresh water: shrimp, waterlouse, crayfish<br>On land: woodlouse | |
| 17 Centipedes and millepedes | Land animals with many legs<br>a Centipedes<br>  1 one pair of legs per segment<br>  2 prominent feelers<br>  3 feed mainly on other animals<br>b Millepedes<br>  1 two pairs of legs per segment<br>  2 feelers not prominent<br>  3 feed on plants | Observe legs along whole length of body<br>Observe on centipedes<br>1 one pair of legs per segment<br>2 prominent feelers<br>Observe on millepedes<br>1 two pairs of legs per segment<br>2 feelers not prominent | Keep in suitable containers |
| 18 Weather. Three contributory factors | a Sun<br>b Air<br>c Water | Observe<br>a different speeds of moving air<br>b water becomes water vapour | Controlled experiments to show that both heat and moving air help water to become water vapour |
| 19 Stars, planets and satellites | a Stars are made of glowing gases. The sun is a star<br>b A planet goes round a star<br>c A satellite goes round a planet | Observe star patterns<br>Collect up-to-date information on artificial satellites | Act the parts of moon going round earth, and earth going round sun. Demonstrate sizes of first four planets and their satellites by scale models |
| (C) General Subjects<br>1 The four main needs of living things | a Oxygen<br><br><br><br><br>b Food<br><br><br><br><br><br><br><br><br><br>c To grow<br><br><br>d To have young | Observe how certain animals obtain oxygen by muscular movements, i.e. by breathing<br>Observe the kinds of foods used by different animals and plants, i.e. living animals or plant foods, dead animal or plant foods, never-alive solids and liquids<br>Observe evidence of growth in animals and plants<br>Collect fruits and seeds, cones and seeds, spore cases, egg cases, etc | |

| Subject | Topic | Collection of Material by Children, and Local Observation | Demonstration and Experiment |
|---|---|---|---|
| 2 Care of living things | Insects need<br>a oxygen<br>b food | | Set up suitable containers for caterpillars and other insects |
| 3 Different stems and roots on herbs trees and shrubs | a Stems have buds and leaves. Most grow towards the light<br>Some plants have:<br>  1 underground stems<br><br>  2 creeping stems<br><br>  3 climbing stems<br><br><br><br><br><br>b Roots have branch roots and root hairs. Most grow away from the light. Some plants have aerial roots | <br><br><br><br>E.g. couch grass, mint, horsetail, corms, potato<br>E.g. strawberry, creeping jenny, iris<br>a twining, e.g. bindweed, honeysuckle<br>b with tendrils, e.g. virginia creeper, pea<br>c with hooks, e.g. roses, brambles<br><br>Observe root hairs<br><br><br>Observe adventitious roots, e.g. ivy | |
| 4 Forcing things to move | Animals alone can force themselves to move<br>a in water<br>b on land<br>c in the air | Observe<br>a different methods of movement<br>b different limbs | |
| 5 Other things are forced to move | By<br>a 1 moving animals<br>  2 moving gases<br>  3 moving liquids<br><br>b gravity<br><br><br><br>c magnetism | Observe<br>a how animals move things<br>b things moved by<br>  1 wind<br>  2 water<br>Observe that things fall towards the earth<br><br>Observe shapes of different magnets | Demonstrate that air and water are in turn forced to move<br><br><br><br>Demonstrate<br>a downhill is easier than uphill<br>b principle of plumb bob<br>Demonstrate<br>a ordinary magnets affect only certain things – chiefly iron and steel<br>b pull is strongest at the ends<br>c things made of iron and steel can be made into magnets |

| Subject | Topic | Collection of Material by Children, and Local Observation | Demonstration and Experiment |
|---|---|---|---|
| 6 We force other things to move on land | By<br>1 carrying<br>2 sliding<br>3 rolling | Observe use of these three methods<br>Observe shapes of 'round' things<br>1 like a drum (cylindrical)<br>2 like a ball (spherical) | Demonstrate<br>a sliding a heavy weight is easier than carrying it<br>b it is easier to slide over a smooth surface than a rough one<br>c for round things, rolling is easiest of all |
| 7 Using rollers and wheels for things which are not round | a Early rollers<br>b Light rollers are better than heavy rollers<br>c Wheels on an axle | Observe<br>a number of wheels on vehicles<br>b how wheels are fixed on toy motor cars, etc | Demonstrate that it is easier to force a heavy load to move on rollers or wheels, than it is to slide it |
| 8 Never-alive mixtures | a Solids<br><br>b Liquids<br><br>c Gases<br>d Solid and gas<br>e Solid, liquid and gas<br>f Liquid and solid<br>g Liquid and gas<br>h Solutions<br>1 solid in a liquid<br>2 liquid in a liquid | Broken concrete, ores, etc<br><br><br><br><br>Smoke<br>Fog | Mix sand and salt<br><br>Make an emulsion of oil and water<br><br><br><br>Make a paste<br>Make a foam<br><br>Sugar or potassium permanganate in water<br>Mix Dettol and water; oil and turpentine |
| 9 Water | a As a food for all living things<br>b Some animals drink it<br>c Wells and reservoirs | | Demonstrate<br>a difference between porous and impervious rocks<br>b how water seeps into a well<br>c purifying water by filtering |
| 10 Salt | a Dissolves in water<br>b Is carried to the sea by streams and rivers<br>c Remains in the sea<br>d Our salt comes from sea water or salt mines | | Demonstrate that<br>a salt dissolves in water; can be tasted<br>b rainwater dissolves salt in the ground<br>c when water evaporates salt is left |
| 11 Oxygen gas | a Present in air and water<br>b Lungs help to take it | Observe<br>a fresh air needed for breathing | Controlled experiments to show<br>a oxygen is used dur- |

| Subject | Topic | Collection of Material by Children, and Local Observation | Demonstration and Experiment |
|---|---|---|---|
| 11 Oxygen gas (cont.) | from the air<br>c Gills help to take it from the water<br>d Used during burning | b fresh air needed for burning | ing breathing<br>b most things cannot continue to burn without oxygen |
| 12 Freezing and melting | a When water vapour freezes it forms snow or frost<br>b When water freezes it forms ice or frost<br>c Solid forms melt into liquid<br>d Melting is not the same as dissolving | Observe frost on window panes and inside refrigerators<br>Observe that heat causes melting | Experiment to find that ice or snow take up more room than water |
| 13 Dead things are used by living things | a Mainly for food<br>b By certain animals for protection against enemies and weather<br>c Human beings alone use dead things in other ways | Observe fungi feeding on cloth, leather, cork, etc<br>Observe nesting material used by birds, hamsters, etc<br>Observe additional uses to which we put dead things, e.g. fuel, clothing | |
| 14 Rock is worn down | a By action of moving water<br>b By action of moving air<br>c Particles carried from place to place by moving water and air | Observe different sizes of sands, gravels, pebbles and stones<br>Observe results of sea action on a rocky shore<br>Observe sand carried by the wind | Display pebbles in shallow dish of water |

## Keeping Specimens for Observation Purposes

### Alive Specimens

As the four main needs of living things are oxygen, food, to grow, and to have young, the two major requirements to be considered when keeping living specimens are invariably oxygen and food. Those from a warmer climate than our own need to be kept in surroundings with a higher temperature than our own. For example, tropical fish and tropical aquatic plants need water maintained by means of a heater and thermostat at a temperature of about 24° Celsius.

Land animals and plants obtain their oxygen from the air, and, as air is in plentiful supply, the major consideration for terrestrial specimens is obviously food. Plants, of course, obtain their food from the water which they acquire, so that in general living animals are more difficult to keep. This applies particularly to those which feed on other living animals. For example, a grass snake is not an easy pet to cater for, because it feeds on living fish or living amphibians; on the other hand, stick insects are very easy to keep because they feed on privet or ivy leaves which – even in a city – are easily obtained.

Assuming the surrounding temperature is suitable, then for aquatic specimens both oxygen and food requirements have to be catered for. Thus for goldfish, tropical fish and tadpoles, the main needs are:

1  Water with a sufficient surface area to permit the diffusion of all the oxygen consumed by the occupants. (The so-called oxygenating plants – underwater herbs – will not in themselves be sufficient in number to replenish all the oxygen absorbed by the animal occupants of the average aquarium.)
2  A sufficiency of the right kind of food, but not an excess. An excess of food in an aquarium leads to pollution – unless of course the food itself is alive.

### Dead and Never-alive Specimens

Dead and never-alive specimens which have been collected for illustration purposes may be considered under the following headings:

20

A *Specimens which may be retained as they are without decomposing, e.g.*

1 *a* Sturdy never-alive specimens such as hard rocks, sea shells and metals etc. Being never-alive, these are not subject to attack by bacteria and fungi.

*b* Sturdy dead parts such as crocodile skin, piece of fur, bone, tree bark, peat, part of a coconut fruit, etc., which are normally too dry for bacteria and fungi to feed on, and which should remain so if kept in dry surroundings.

2 Certain flowers which are amenable to drying. Examples are hydrangea flowers and reedmace (sometimes mistakenly termed bulrush). The cut stems of these may be stood in vases or jars without water, and kept in dry surroundings.

Some specimens, *e.g.* the two parts of the shell of a bivalve mollusc, part of the skin from a dead hedgehog, part of a crab shell or claw etc., may be acquired with some of the unwanted remains of the dead animal still attached. These may be cleaned and disinfected in the following way:

1 Remove any unwanted remains.
2 Immerse the specimen in a mixture of hot water, Dettol and detergent.
3 After a day or so, remove the specimen from the mixture and allow to dry.

*Note*
The two parts of the shell of a bivalve mollusc may, after drying, be used to show the appearance of the complete animal. Simply glue the edges together. They need of course to be from the same mollusc. Separate parts from two molluscs, *e.g.* two cockles, will not fit together properly.

B *Specimens which may be retained as they are without decomposing, but which need protection against loss or against damage by careless fingers, e.g.*

A small piece of soft valued rock, some sand from a particular beach, a small fossil, the empty shell of a bird's egg from someone's old collection, etc. Such specimens can be safeguarded by keeping them in suitable containers, *e.g.*

21

1   small screw-topped jars
2   Alka Seltzer or similar tubes
3   specimen tubes (fitted with either plastic caps or corks)
4   glass-topped specimen boxes.

1 and 2 can be supplied by children, and 3 and 4 obtained from suppliers of biological equipment. A lining of cotton wool is often useful, and suitable, but meaningful labels made out of coloured card (red for animal, green for plant, and yellow for never-alive) help to furnish a collection of such containers with system and colour.

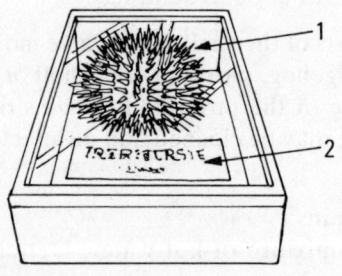

1 Specimen, e.g. small sea urchin with spines.
2 Label (red) with words: 'Simple Animals with Spiny Skins
    A SEA URCHIN
Robin Hood's Bay, Yorks., 1976.'
GLASS-TOPPED SPECIMEN BOX

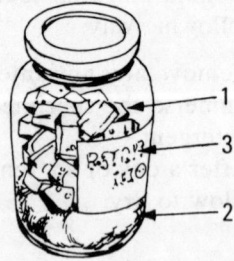

1 Specimen, e.g. Blue John Ore.
2 Cotton wool.
3 Label (yellow) with words: 'Rocks, Minerals and Ores
    CRYSTALS OF BLUE JOHN
Castleton, Derbyshire, 1976.'
    SCREW-TOPPED JAR

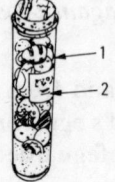

1 Specimen, e.g. a casual beach profile, showing pebbles and sand.
2 Label (yellow) with words:
'ROCK from the beach at Bognor Regis, Sussex, 1977'.
SPECIMEN TUBE WITH CORK
    OR PLASTIC CAP

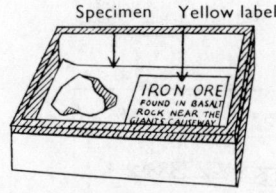

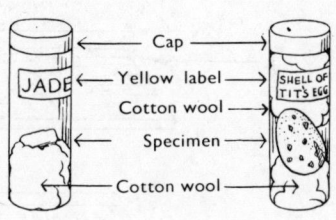

PROTECTION FROM LOSS
OR CARELESS FINGERS

**Method for Retaining Selected Specimens such as Feathers, Section of Skin**

Use the following materials as shown in the diagram on p.24:

1 piece of hardboard, thin wood, or stiff cardboard
2 thin coloured card – red, green, or yellow according to whether specimen is animal, plant, or never-alive
3 specimen
4 self-adhesive plastic
5 adhesive tape

C *Specimens which need protection from the bacteria and fungi which cause decay, e.g. most parts of animals and plants*

*Reasonably flat specimens, e.g.* autumn leaves, flower petals, etc. These may be preserved from attack by bacteria and fungi by sealing them from the air. The following method has been found

23

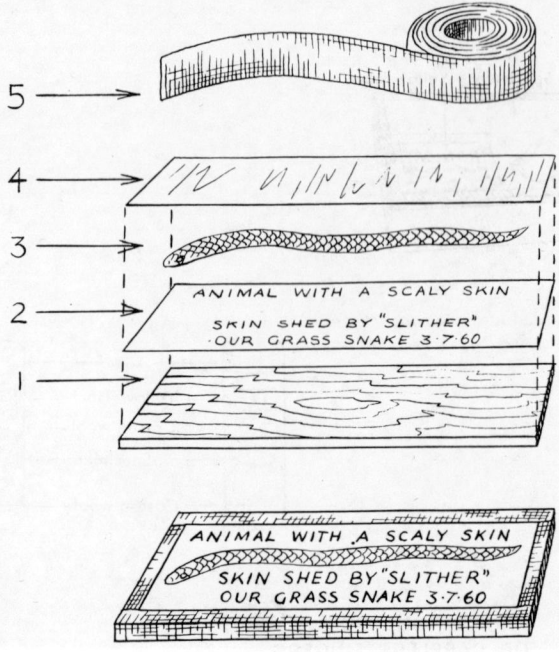

suitable for mounting such specimens in notebooks, or – for more permanent collection – on card, glass or perspex.

*a* Press the specimen first. This tends to reduce the water content, and make the specimen too dry for bacteria and fungi to feed on.

*b* Place in position.

*c* Fix in position with strips of cellulose tape (or self-adhesive plastic sheet).

*d* Add a suitable label of coloured card (red for animal, green for plant).

*Notes*

*i* Providing the tape or self-adhesive plastic sheet completely covers the specimen, it should provide a transparent air-tight seal, and for the purpose of temporary preservation in notebooks, the specimens need not actually be pressed first.

*ii* The disadvantage of ordinary cellulose tape is that it eventually shrinks, so that for a permanent mount it is advisable to use self-adhesive plastic.

24

**Method for Obtaining a Dry Mount of a Specimen on a Sheet of Glass or Perspex**

This method is suitable for large flat specimens where it is desirable to have both sides on view, *e.g.* a fern leaf showing spore cases on the undersides of the leaflets.

1  Place in position on a length of glass or perspex
   *a* the pressed specimen to be mounted
   *b* coloured card bearing information.
2  Fix these in position with a suitably-sized sheet of self-adhesive plastic. This not only fixes the specimen and card in position, but provides at the same time a transparent air-tight seal. The plastic should extend the full width and length of the glass or perspex, and be turned over all four edges.
3  Finally run a strip of adhesive binding tape round all four edges of the glass or perspex to prevent them coming into contact with delicate fingers.

*Notes*
 *i* Perspex is of course much less fragile for children to handle than glass. It is also more expensive.
*ii* A formica or similar surface provides a good working base when using self-adhesive plastic or cellulose tape as – if any part of these materials adhere to it – they can be peeled off without removing any paint or varnish.

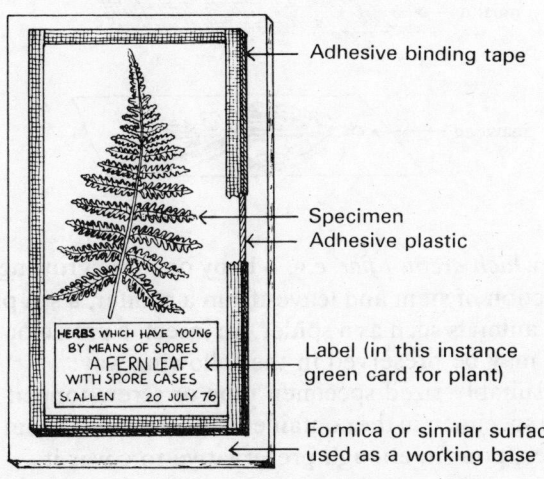

HERBS WHICH HAVE YOUNG
BY MEANS OF SPORES
A FERN LEAF
WITH SPORE CASES
S. ALLEN    20 JULY 76

Adhesive binding tape

Specimen
Adhesive plastic

Label (in this instance green card for plant)

Formica or similar surface used as a working base

25

## Method for Obtaining Dry Mounts of Sea Plant

1 Float specimen in water.
2 Slide a sheet of card underneath, and lift.
3 After the water has drained off, tease out any fronds.
4 Cover with a layer of butter muslin.
5 Cover this with blotting paper.
6 Cover this with several layers of newspaper.
7 Leave under pressure for several days.
8 Remove gently from the card, and mount in the same way as the reptile skin on page 24 or the fern leaf on page 25.

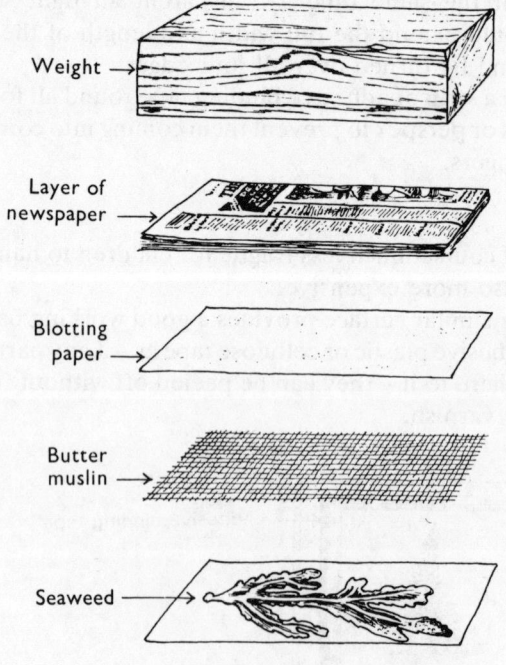

Weight →

Layer of newspaper →

Blotting paper →

Butter muslin →

Seaweed →

*Specimens which are not flat*, *e.g.* a baby oak tree growing from an acorn, a section of stem and leaves from a conifer, a sea plant, and small dead animals such as a spider, an insect, a prawn or a shrimp etc. These may be preserved in the following way:
*a* Select a suitably-sized specimen tube or screw-topped jar.
*b* Place the specimen in the container and, after rinsing out gently in cold water, pour in enough preservative to cover it.

*c* Leave for a few days, and then if some discolouration has taken place, rinse out and fill up with fresh clear preservative.

*d* Add a suitable label made out of coloured card (red for animal, green for plant).

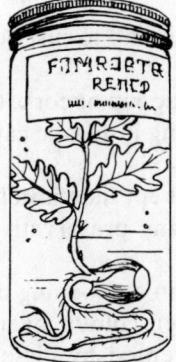

1 Plastic cap
2 Label (in this case green for plants) with words

> Plants with roots, stem and leaves
> **A BABY OAK TREE**
> 5% form. +                    17.6.76.

3 Liquid preservative
4 Specimen
The + sign indicates that the formaldehyde contains Copper Acetate to retain the green colour of the leaves.

1 Plastic cap
2 Label (in this case red for animals) with words

> Animals with Crusty Skins
> **A SEA SHRIMP**
> 5% form.                    25.7.75

3 Liquid preservative
4 Specimen

PROTECTION FROM DECAY (i.e. attack by fungi or bacteria)

*Notes*

*a* A good all-round preservative is 5% or 10% formaldehyde. This can be obtained from biological suppliers, usually as a 40% solution, known as formalin. A 10% solution can be made by mixing three parts of water to one part of 40% formaldehyde. Mixing the 10% solution with an equal amount of water gives a 5% solution which has been found to be strong enough to prohibit the development of bacteria and fungi in small specimens without affecting the tissue.

Experiments over the last few years have shown that a 5% solution is suitable for almost all the preserved plant and animal parts used during science lessons with juniors.

b It is better to err on the side of a weaker solution than on the side of a stronger one.

c Before fixing the label, fill up with preservative:

   i in a screw-topped jar to the rim

   ii in a specimen tube with a cork or plastic cap to just below the rim.

d In the case of a specimen tube with a plastic cap or cork, twist the cap or cork into position instead of pressing it. Plastic caps make neater and better tops than corks.

e Specimens preserved in this way may be kept indefinitely with a very occasional 'topping-up' to replace any preservative which has evaporated.

f i Some green plant parts, e.g. new leaves in an opening bud tend to lose chlorophyll, and thus their green colour, even in 5% formaldehyde. This can often be prevented by dissolving as much copper acetate as possible in the 40% formaldehyde before diluting it.

   ii Clearer solutions of formaldehyde are obtained if the 40% mixture is passed through a filter before mixing with water.

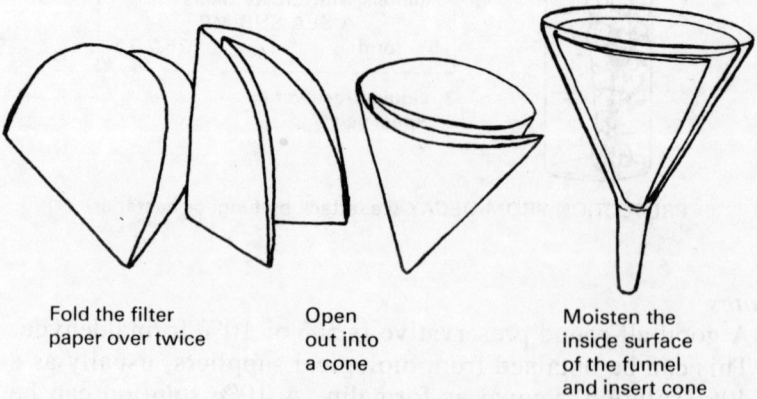

Fold the filter
paper over twice

Open
out into
a cone

Moisten the
inside surface
of the funnel
and insert cone

PREPARATION OF FILTER PAPER AND PLASTIC FUNNEL
FOR FILTERING 40% FORMALDEHYDE

A few specimens which are not flat may be preserved in other ways, e.g.

*a* A starfish which it is desired to preserve may not always fit into a screw-topped jar. Another way is to fix and preserve the starfish in a shallow vessel containing 10% formaldehyde. Leave completely immersed for a few days to allow the formaldehyde to penetrate the soft inner tissue, and then remove and allow to dry. The spiny skin of the animal should then be sufficient protection.

*b* A complete flower may be preserved in a tin of sharp dry sand or fuller's earth – a method of pressing without flattening. Almost empty the tin first and push the flower stalk into the remaining sand. Pour the remainder of the sand carefully around and between the flower parts. Leave for about three weeks in a very dry place, to allow the sand to reduce the water content of the specimen so that it is too dry for bacteria and fungi to feed on. Finally remove the flower and place dry in a screw-topped jar to protect it from careless fingers. The inclusion of a few crystals of paradiChlorbenzene, or some mothball fragments will serve as protection against possible invasion by mites.

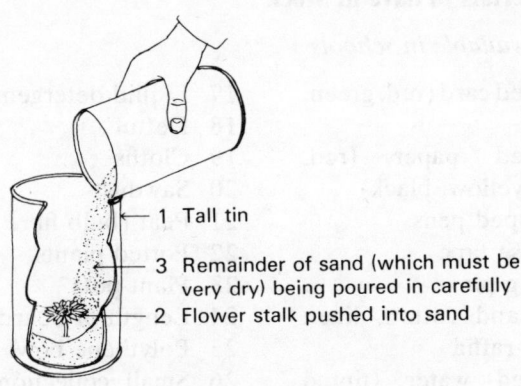

1 Tall tin

3 Remainder of sand (which must be very dry) being poured in carefully

2 Flower stalk pushed into sand

PRESSING WITHOUT FLATTENING

## Killing Specimens for Preservation Purposes

In the living world in general, there are two main reasons for killing – for food and for self-protection against enemies.

Probably the only other motive with a claim to validity is that of killing for the purpose of scientific study, in order that something of value to human beings may be learnt. Even so, it is obvious that every effort should be made to avoid indiscriminate killing.

When any living plant or animal is immersed in preservative, *e.g.* formaldehyde or alcohol, the preservative itself acts as a killing agent. However, if it is desired to kill a specimen before preserving, then the following equipment is simple and satisfactory:

1  jam jar and tin lid
2  ink bottle containing cotton wool
3  chloroform.

*Method*

1  Pour chloroform into the ink bottle containing cotton wool.
2  Invert jam jar containing specimen over ink bottle, and leave until killing is complete. The tin lid provides a smooth surface on which to place bottle and jar.

**Useful Materials to have in Stock**

*Normally available in schools*

1  Coloured card (red, green, yellow)
2  Coloured paper (red, green, yellow, black)
3  Felt-tipped pens
4  Cellulose tape
5  Drawing pins
6  String and cotton, linen thread, raffia
7  Coloured water (tinted with red or blue ink)
8  Blotting paper
9  Gummed labels
10  Gummed paper strip
11  Coloured tape
12  Cotton wool
13  Modelling clay
14  Tennis balls
15  Elastic bands
16  Paper clips
17  Liquid detergent
18  Dettol
19  Cloths
20  Sawdust
21  Peat (bulb fibre will do)
22  Potted plants
23  Plant pots
24  Length of board
25  Polythene bowl
26  Small collection of oddments which a magnet will attract
27  Small collection of oddments which a magnet will not attract
28  Steel sewing and darning needles
29  Steel sewing pins
30  Hammer
31  Paper tissues

## Normally provided by children

1 Large jars, *e.g.* toffee jars
2 Sundry clear glass jars with screw tops
3 Jam jars
4 Small ink bottles
5 Medicine bottles
6 Alka Seltzer tubes
7 Table tennis balls
8 Candles
9 Paste powder, flour or starch
10 Oil, *e.g.* machine oil
11 Sandpaper
12 Common salt, and sugar
13 Common sand
14 Broken pebbles
15 Piece of broken concrete or conglomerate rock
16 Specimens of ore
17 Collection of samples of different grades of sand and gravels
18 Tall cylindrical tin with lid
19 Length of broom handle or dowel rod for rollers
20 Non-mechanical toys with wheels and axles
21 'Plumb-bob' *e.g.* weight or conker suspended on a thread
22 Mirror
23 Lead strip for weighing down aquatic plants
24 Magnets of sundry shapes and sizes
25 Small panel pins or tin tacks
26 Small piece of plastic sponge
27 A brick
28 Dried peas, butter beans or other seeds
29 A bird's old nest
30 Pine or other cone with winged seeds
31 Coral
32 Egg cases, *e.g.* dogfish, skate, whelk
33 Section of reptile skin showing scales
34 Dead part of a crustacean, *e.g.* claw, or part of crusty skin of crab or lobster
35 Tobacco tin or similar
36 Length of flat elastic
37 Expanded polystyrene
38 Small polythene bags
39 Plastic detergent bottles
40 Matchboxes
41 Circular biscuit or cake tin (for insect cages)
42 Plastic boxes
43 Corks
44 Tin lids, various
45 Glass marble
46 Golf ball
47 'Empty' bottles containing lingering smells

## Normally for purchase

1 Methylated spirits
2 Paraffin

3 Turpentine
4 Aquarium gravel
5 Stiff acetate sheet for insect cages
6 Adhesive tape for above
7 Polythene funnel
8 Good quality magnets
9 Potassium permanganate
10 Magnifying glass

*Useful extras*

1 An aquarium or an old sink containing aquarium gravel
2 Glass or plastic cover for above
3 Air pump for aerating water, together with rubber or plastic air tubing, screw clamp and diffuser stone
4 Vivarium for amphibians or reptiles
5 Stick insects, or eggs of such

**Some Useful Preserved Specimens**
(See Keeping specimens for observation purposes)
1 Any small complete herb, tree or shrub to show roots, stems and leaves, *e.g.* a small oak tree growing from an acorn
2 Various fungi, *e.g.* mushrooms, toadstools, moulds, mildews, tree fungus
3 A section of stem with adventitious roots, *e.g.* ivy
4 Stem and spore cases of horsetail
5 Section of stem to show a simple leaf without a stalk
6 Section of stem to show a simple leaf with a stalk
7 Section of stem to show a compound leaf with end leaflets
8 Section of stem to show a compound leaf with side leaflets
9 Section of twining stem with host, *e.g.* bindweed
10 Section of stem with hooks, *e.g.* rose
11 Section of stem with tendrils, *e.g.* pea
12 Insect or other eggs on plant parts
13 A small lizard
14 Any other small reptile, *e.g.* terrapin
15 Eggs, larvae and pupae of moths or butterflies
16 Eggs, larvae and pupae of two-winged flies, *e.g.* bluebottle
17 Spider – showing eight legs clearly
18 Harvestman
19 Woodlouse
20 Centipede
21 Millipede
22 Sea shrimp or prawn

23 Starfish (if possible with one new ray growing)
24 Fish killed by fungus with fungus apparent
25 Great water moss (Fontinalis)
26 Any other small freshwater plant
27 Sea-anemone
28 Sea-mat
29 Small crab
30 Life history of an insect with a three-stage life
31 Life history of an insect with a four-stage life
32 Life history of amphibian, *e.g.* frog, toad, newt

## Some Useful Mounted Specimens

1 Fern leaf to show spore cases
2 Simple and compound leaves
3 Adult moths and butterflies
4 Adult two-winged flies
5 Life history of an insect with a three-stage life
6 Life history of an insect with a four-stage life

*Note*
Mounted or preserved specimens can be bought from biological suppliers. When ordering them it is wise to state exactly which points are to show clearly and to ask for English terminology.

## General Materials for Mounting and Preserving

1 Specimen tubes of various sizes with plastic caps. The most useful sizes are: 75 mm × 19 mm, 75 mm × 25 mm, 100 mm × 25 mm, 125 mm × 25 mm, 150 mm × 25 mm
2 Clear glass jars with screw tops
3 Alka Seltzer tubes
4 10% formaldehyde. Diluted from 40% formaldehyde (formalin)
5 Copper acetate (for retaining green colour in plants)
6 Glass-topped specimen boxes
7 Hardboard or stiff cardboard, or thin boxwood
8 Rectangles of perspex or clear glass
9 Self-adhesive plastic sheeting
10 Cellulose tape
11 Dettol

12  Detergent
13  Strong glue
14  Para-diChlorbenzene (or bits of mothballs)
15  Butter muslin
16  Coloured card (red, green, yellow)

## The Use of the Science Table

The science table helps children to look at things systematically. In its layout it should reflect the fact that science, very simply, is concerned with

1   Things in general
    *a* Living things
    *b* Dead things
    *c* Never-alive things
2   What happens to them, or 'happenings'

Children who have worked through Pupils' Book 1 of this series will be familiar with the division of their science table into sections marked *alive, dead* and *never-alive*. They will also be aware of the colour system used throughout the series –
    Red for animal (whether living or dead)
    Green for plant (whether living or dead)
    Yellow for never alive things (whether solid, liquid or gas)
Their science table will have had a covering of coloured papers, like this

| ALIVE | DEAD | NEVER ALIVE |
|---|---|---|
| Red (*i.e.* living animal) | Red (*i.e.* dead animal) | Yellow (*i.e.* solid, liquid or gas) |
| Green (*i.e.* living plant) | Green (*i.e.* dead plant) | |

or the names of exhibits will at least have been written on card of the appropriate colour.

This year the never-alive (yellow) section should be divided into three – solid, liquid, and gas.

The word 'table' is used, of course, only as a convenient description; several shelves and the top of a cupboard and a window ledge might together act as the science table.

## Alive Section

Anything is alive so long as it continues to respire, but respiration is not always noticeable. Fresh fruits, flowers and greens are living plant parts, as are uncooked potatoes and peas. Potatoes are swollen stem parts, and will put out shoots if stored for long enough, as children may have observed in their own homes. Dried peas, being seeds, will retain life for years if stored in a container which allows them sufficient fresh air for their oxygen requirements. Dating the container in which pea seeds are kept will enable children to appreciate, when the time comes to germinate some, just how long seeds are able to retain life.

The observable difference between non-microscopic forms of animal and plant life is that living animals are capable of movement from one place to another of their own choosing, whereas plants are not. This voluntary movement is quite distinct from the movement of plants in the wind or towards the light. It is also different from the movement of never-alive things activated by some independent force, for example, an aeroplane or motor car, or wind and sea and rivers, or a dislodged rock rolling down a hill.

Typical specimens for the *living animal* section could include pet hamster, fish, frogspawn, tadpoles, and terrestrial or aquatic insects in appropriate containers.

Typical specimens for the *living plant* section could include bulbs, carrot tops, twigs, beans, peas, acorns, and various other living samples of roots, stems, leaves, flowers, fruits, seeds; and specimens of simple plants, which of course have none of these parts.

## Dead Section

Things to be included in the dead section should consist entirely of dead material (*i.e.* cellular in structure). These would include not only whole animals and plants, but parts of animals and plants which are dead. (For mounting and preserving see Introduction.) Because an object is only part of an animal or plant, it is not necessarily dead. Fresh fruits and flowers are only parts of plants, but they are alive, although, of course, they may be slowly dying. The top of a carrot, a section of potato with an 'eye' in it, and a cutting from a plant stem may all be capable of developing into complete new plants if given suitable conditions. Boiling, of course, kills most living things, although microscopic forms such as the amoeba may withstand temperatures of 100°C or more. Nevertheless, the parts of animals and plants which have been cooked may be considered dead.

Some manufactured articles are made entirely of dead material, *e.g.* wool, fur, leather, or wood. In fact the human mammal relies a great deal on the dead parts of other animals and plants to furnish the material for food, clothing and even shelter.

Typical specimens for the *dead animal* section could include any mounted or preserved specimens of whole animals, as well as parts of animals such as hair, fur, feathers, leather, snakeskin, teeth and bones.

Typical specimens for the *dead plant* section could include any mounted or preserved specimens of whole plants, or whole plant parts, together with such plant material as wood, cork, bark, autumn leaves, peat, coal.

## Never-Alive Section

Never-alive things may occur in various ways.

1 Some occur naturally, such as air, water, rock. These are our chief concern.

2 Some are excreted by, or extracted from, living or dead things, *e.g.* sugar, milk, fat, oil, and resin (which is the basis of turpentine). Perspiration and tears are obvious never-alive excretions.

3 Some are manufactured by living animals. For example, glass, jewellery and metal goods are made by humans; a spider's web, a silken cocoon, and the shell of a bird's egg are manufactured by other animals. A mollusc, such as the whelk, cockle or snail, may be alive, but the limestone shell which it builds on the outside of its body, is as never-alive as the bricks and mortar of a human home.

Pedantically speaking, every article manufactured by an animal is never-alive, in the sense that it has been made in that shape and size and has not grown from a smaller version. However, where an object consists entirely of dead material, then logically it is dead material. A wooden peg; for example, has not lived, respired, fed and grown *as a peg*, but it consists entirely of material which has lived, respired, fed, grown and died. On the other hand, a metal shovel with a wooden handle is an object which has been manufactured from a combination of never-alive and dead material.

36

# INTRODUCTION

It may be considered expedient to leave grouping in the never-alive section until the first lesson, when the three forms of never-alive things – solid, liquid and gas – are dealt with.

# THE THREE FORMS OF NEVER-ALIVE THINGS

## SOLID, LIQUID, AND GAS

**Demonstration Material**

As many as possible of the following:

1  Yellow card or yellow paper for the *never-alive* section of the science table.
2  Various solids, *e.g.* modelling clay, piece of brick, india-rubber, elastic band, metals, rocks, sea shells, plastic, polystyrene, glass.
3  Various liquids in small bottles, *e.g.* water, ink, methylated spirit, oil, paraffin, turpentine, Dettol, detergent, etc.
4  Various 'empty' containers in which noticeable odours remain, *e.g.* bottles once containing perfume, pickled onion, vinegar, Dettol, etc.
5  Coloured water, and three or four transparent containers of different shapes and sizes, *e.g.* small milk bottle, medicine bottle, glass jar, plastic box.
6  Tin lid, screw-topped jar with lid, paper and possibly a little machine oil, matches.

**Sample Link Questions**

1  What are the three kinds of things in the world? (*Alive, dead, and never alive*)
2  What are the two kinds of living things? (*Animals and plants*)
3  What are the two kinds of dead things? (*Animals and plants*)
4  What is the common name for sand, sandstone, stones and pebbles? (*Rock*)
5  What does water turn into when it freezes? (*Ice*)
6  Is air alive, dead, or never alive? (*Never alive*)
7  Can we see air? (*No*)
8  Why do we keep ink and milk in bottles? (*To stop them flowing away*)

## Relevant Information

Lessons in Book 1 showed that:

1   the three main kinds of things in the world are *alive, dead* and
*never alive*;
2   the two main kinds of living things are *animals* and *plants*;
3   the two main kinds of dead things are *animals* and *plants*.

Living and dead things are always solid in form. Never-alive things
are encountered in one of three forms. The purpose of this lesson is:

1   to show that the three forms in which never-alive things are
found are –
*a* solid
*b* liquid
*c* gas
2   to show the main differences between these three forms.

The fact that some never-alive things, *e.g.* water, are capable of
existing in more than one of these three forms, depending on
temperature, is the subject of a separate lesson in Book 3.

### *a Solid*

A substance which is in solid form has a definite size, *i.e.* volume,
and can maintain a fixed shape. This shape can be changed only by
the application of some force, *e.g.* a piece of lead or modelling clay
can be bent, aluminium foil or a polythene bag can be crumpled, and
cast iron or a strip of polystyrene can be broken. Rubber and plastic
objects may be twisted out of shape, but they tend to return to their
original shape. Most solids have colour but some, such as clear glass
and clear plastic are colourless. Some solid things, *e.g.* a tin or a
bottle may be hollow inside, but the metal or the glass having size
and shape, are of course still solid.

   Associated with solids are the characteristics of tenacity, ductil-
ity, malleability, hardness, brittleness and elasticity.

### *b Liquid*

A quantity of a substance in liquid form has a definite size or
volume, but no fixed shape. This lack of shape results in a liquid
adopting, so far as is possible, the shape of any solid vessel in which

it is contained. There is one metal – mercury – which obeys these definitions at ordinary temperatures, and is therefore in the liquid form at temperatures at which other metals are in the solid form. A characteristic of a liquid is that when it is completely at rest, its surface is horizontal and level, and never slanting. Many liquids have colour, but some, *e.g.* water, are colourless.

At a given temperature, some liquids are more viscous than others. That is to say they are nearer to the solid state than others. Treacle, for example, is more viscous than oil, and engine oil is more viscous than water.

### c Gas

A substance which is in the form of a gas has no definite size or volume, and has no fixed shape. Thus, in the gaseous state, a substance always tends to spread throughout the available space in any vessel which contains it, in addition to taking the shape of its container. Most gases are colourless, and cannot be seen, but a few kinds, *e.g.* chlorine gas, have colour.

A gas is easily compressed, whereas a liquid is almost incompressible. The illustration of the skin-diver's aqualung in the Pupil's Book shows a cylinder of air which has been compressed.

*Air* is of course the commonest example of gas. It is a mixture of gases – approximately four-fifths nitrogen, and one-fifth oxygen, with traces of other gases such as water vapour and carbon dioxide.

*Coal gas* is a mixture of gases. It consists mainly of methane, carbon monoxide and hydrogen, and includes traces of nitrogen, carbon dioxide and oxygen. The containers used for storing coal gas are correctly termed gas holders. At one time coal gas containers served the dual purpose of storing and measuring and were thus known as gasometers. It is because of this that the term *gasometer* is still sometimes mistakenly applied to present-day gas holders.

*Smoke,* having neither size nor shape, has the form of a gas, and is perhaps the best visible example of how the gaseous form differs from the solid and the liquid forms. Smoke is not made up entirely of gas however. It is a mixture consisting of a suspension of fine particles of solid in a gas. The smoke from a coal or wood fire, for example, is a mixture of fine particles of solid carbon (soot) together with carbon dioxide gas and water vapour. Thus, although a smoke demonstrates the form that a gas takes, what we actually see in a smoke is not the gas itself, but the particles of solid which it carries.

*Vapour.* Below a certain temperature, known as the critical temperature, a gas may be described as a vapour. Each gas has its own critical temperature. At temperatures *below* the critical temperature a gas can be liquefied by

   *a* lowering the temperature
*or b* increasing the pressure.

*Water vapour and steam.* The critical temperature for water in the gaseous form is about 374°C. Thus, when water evaporates and becomes a gas, it is termed *water vapour*. When water boils at 100°C and normal pressure, the resulting gas still comes under the category of vapour but is termed *steam*. This, like most gases, is invisible, and is present in the space between the spout of a kettle and the place where it begins to condense in the form of a cloud of tiny drops of water. It is this cloud of condensed vapour which is often wrongly referred to as steam. It should of course be called condensed steam, as it is made up of tiny drops of water, like a cloud in the sky or mist near the ground.

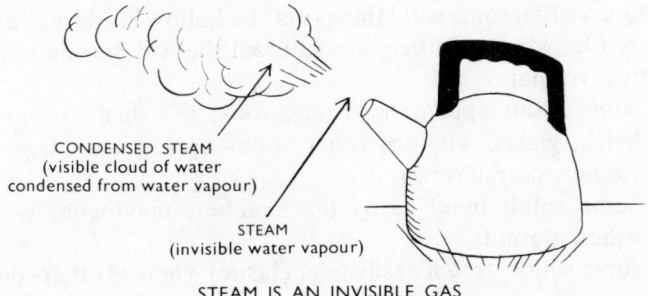

CONDENSED STEAM
(visible cloud of water
condensed from water vapour)

STEAM
(invisible water vapour)

STEAM IS AN INVISIBLE GAS

*Fumes.* The term *fume* is applied to a gas or smoke which is unpleasantly perceptible to our sense of smell. When we sense any substance by means of smell, it is due to part of that substance in the form of gas entering our nostrils and stimulating the nerve endings within.

*Notes*

1 A *fluid* has no shape to prevent it from flowing. Consequently the term can apply to either a liquid or a gas.

2 A liquid will give off tiny particles of its substance (molecules) in the form of gas (or vapour), so that it can be converted into a gas without necessarily reaching its boiling point. This is what happens to water in a puddle on a hot day. This evaporation or 'drying-up' takes place at the surface, and therefore the greater the surface area, the greater the rate of evaporation.

3 The shrinkage of dead things is due to the evaporation of liquids which they contained when they were alive.

4 In that they have size and shape, living animals and plants are solid in appearance. They also contain never-alive solids; and as they take in both oxygen and water, they also contain both liquid and gas.

## CODE

### A *For Solids*

1 Observe the definite shapes of various solids, *e.g.* round, flat, oblong, cube, sphere, irregular.

2 Demonstrate that when the size and shape of solids are altered, they retain their new size and shape. For example, break off a piece of brick; cut off a piece of modelling clay and mould it.

3 Observe that some solid things may be hollow inside, *e.g.* a tin or a bottle, whereas others are solid all the way through.

4 Observe that:

    *a* some solids appear hard (*e.g.* rock, sea shells, iron, sand, brick, glass), whereas other solids appear soft, (*e.g.* clay, rubber, polystyrene);

    *b* some solids bend easily (*e.g.* rubber, polythene) whereas others do not;

    *c* some solids stretch easily (*e.g.* elastic) whereas others do not.

### B *For Liquids*

1 Demonstrate that a liquid can keep its size (volume) but not its shape. (See illustration on page 43). Fill a small milk bottle or jar with coloured water and observe:

    *a* that it takes the shape of whatever vessel it is poured into, *i.e.* that it has no shape of its own.

*b* that when it is finally poured back into its original container, it occupies the same space as before, *i.e.* that the size (volume) remains the same.

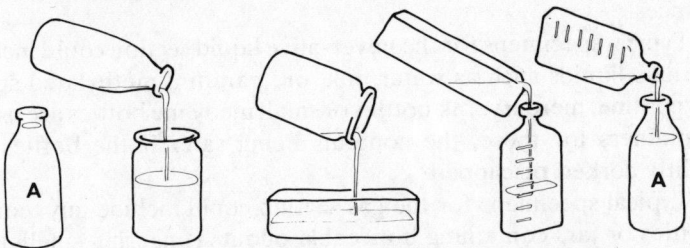

A LIQUID HAS NO SHAPE OF ITS OWN, BUT KEEPS ITS SIZE

2 Demonstrate that when a jar containing a liquid is tilted, the liquid comes to rest with its surface flat and level instead of slanting.
3 Observe that some liquids pour more easily than others, *e.g.* water pours more easily then engine oil, and milk pours more easily than cream.

### C *For Gases*

1 Observe from smoke that a gas has no particular shape, and no particular size. (The gases in the smoke are of course invisible. It is only the solid particles which can be seen.)
2 Collect some smoke for observation purposes:
   *a* light a small piece of paper in a tin lid
   *b* remove the lid from a screw-topped jar and, as the smoke rises, invert the jar over it so as to collect some
   *c* replace the lid so as to trap the smoke.

*Note*
The addition of a *little* oil to the paper before lighting will result in a blacker smoke.

3 Observe that some gases are perceptible to our sense of smell. For example, when we smell liquids such as perfumes or solids such as moth balls, it is due to vapour from these substances entering our nostrils and being sensed by the nerve endings within.

D Collect various never-alive solids, liquids and gases for the science table and divide the *never-alive* section into three parts labelled SOLIDS, LIQUIDS, GASES.

Typical specimens for the *never-alive* solid section could include various metals, salt, glass, plastic, collections of rock including clay and sand, and various articles manufactured from never-alive substances.

Typical specimens for the never-alive liquid section could include various liquids such as water, ink, oil, paraffin, methylated spirit, turpentine, mercury. Ink bottles or small medicine bottles are useful containers for these, the contents being safer if the bottles are tightly corked or capped.

Typical specimens for the gas section could include any 'empty' bottles or jars containing noticeable odours (lingering smells) *e.g.* from perfume, pickles, vinegar, Dettol, etc.

*Note* Clear plastic or clear glass containers with the labels removed could provide material for a simple game, the object of which would be to guess from the lingering smell the name of the original contents.

### Written Work

1  A <u>solid</u> has a size and a shape of its own.
2  A <u>liquid</u> has size, but no shape of its own.
3  A gas has no size or <u>shape</u> of its own.
4  Air is a mixture of <u>gases</u>.

## 2  THE FOUR MAIN NEEDS OF LIVING THINGS

### OXYGEN: FOOD: TO GROW: TO HAVE YOUNG

### Demonstration Material

1  Any examples of living animals or plants.
2  Any examples of dead animals or plants, mounted or preserved, or suitable parts of these.
3  Any examples of seeds.
4  Any examples of eggs or egg cases.

## Sample Link Questions

1  What are the two kinds of dead things? (*Animals and plants*)
2  What are all dead things before they die? (*Alive*)
3  What are the two kinds of living things? (*Animals and plants*)
4  What is the difference between living animals and plants? (*Living animals move from place to place*)
5  What are the three forms of never-alive things? (*Solid, liquid and gas*)
6  What is air a mixture of? (*Gases*)
7  What do fish and some other water animals use for breathing? (*Gills*)
8  What do all living animals and plants need in order to grow? (*Food*)
9  What is one never-alive liquid which all living things use for food? (*Water*)
10  What do all baby mammals feed on? (*Milk*)

## Relevant Information

The fundamental aim of any living species is the propagation of its own kind. Growth to maturity is essential first. In order to grow, both oxygen and food are necessary. It can be said, therefore, that living things differ from both dead and never-alive things in that they have four major needs:

1  oxygen
2  food
3  to grow (to maturity)
4  to have young

The purpose of this lesson is to introduce these as the four main needs of living things.

In Book 1 there were two lessons on feeding. One taught that living things, dead things and never-alive things are used as food. The other was concerned with the food of animals, in that

a some feed on plants
b some feed on other animals
c others feed on both plant and animal foods

The fact that living things need to grow was also dealt with in Book 1.

45

In the present book there are separate lessons on

*a* oxygen gas (No 18)
*b* the methods by which animals have young (No 21)
*c* the methods by which plants have young (No 24)

## Living Things Need Oxygen

To both living plants and living animals, oxygen is perhaps the most important never-alive thing. Both plants and animals require energy in order to utilise food for body-building purposes, and living animals require additional energy to move from place to place. The energy for these and other cellular activities is obtained by both living animals and living plants from the oxidation of foodstuffs.

Oxygen gas is absorbed into both animals and plants as a first step in a process known as *respiration*. In the plant kingdom, and amongst many animals, this absorption results from a straightforward diffusion of oxygen gas from an area of higher concentration outside the living thing to an area of lower concentration within it. Some animals accelerate the process by muscular movements which force a stream of air or water into contact with the body surface through which the absorption takes place. This kind of mechanical assistance is what is meant by *breathing*.

At normal temperatures oxygen is a gas constituting about one-fifth of the atmosphere. It is slightly soluble in water, and it is this dissolved oxygen which is used by fish and other living animals and plants which absorb their oxygen from water. As oxygen is in plentiful supply in the air, those animals and plants which live on the land have little difficulty in obtaining it. Many of the animals and plants which live in fresh water or sea water obtain their oxygen from that which which is dissolved in the water. Other animals come to the surface and obtain their oxygen directly from the air. Some aquatic plants obtain their oxygen from the air by means of leaves raised above the surface.

Anything is alive so long as it continues to respire, although respiration may not always be noticeable. Eggs and seeds are alive, and using oxygen. Some may live a considerable length of time until favoured with suitable conditions for feeding and growing. Dried peas and beans, for example, are seeds which will live for years if stored in containers which permit them sufficient fresh air.

## Living Things Need Food

All living things need food in order to grow bigger, and anything which is used by the body for building purposes may be described as a food. It may be solid or fluid, and may be taken in in various ways, such as by absorbing, drinking or eating. Roots are a development of the plant kingdom to assist in feeding, and mouths are a development of the animal kingdom. Simple plants are without roots, and the simplest animals are without a mouth.

Water is a food which all living things need, and a considerable amount of the bodies of plants and animals is water. It fills and distends body tissues. It is the basis of sap in plants and of blood in those animals which have blood.

Unlike air, food is not always in plentiful supply. In general, most animals have to overcome more difficulties than plants in obtaining their supplies. Green plants – those containing chlorophyll – obtain their food from such never-alive materials as water, dissolved salts, and the carbon from carbon dioxide. The higher plants – those with roots, stems and leaves – feed in this way. Certain classes of simple plants – simple plants being those without true roots, stems and leaves – also feed in this way. Representative of these are the algae, and the moss and liverwort class. On the other hand, fungi and bacteria are two classes of simple plants lacking chlorophyll, and therefore relying for their food on dead things or living things.

This dependence for food on dead things or on other living things is a characteristic of all animals. Some animals feed on plants, and some feed on the animals which feed on the plants. In fact, as there are certain microscopic forms of plant life which are capable of forcing themselves to move, it is probable that the only certain distinction which can be drawn between animals and plants is that no animal is capable of converting inorganic material into organic during feeding. Animals must therefore rely directly or indirectly on food partly made up by plants. This factor has made it necessary for animals to be able to move from one place to another.

It is during the feeding process known as photosynthesis that the green plants (those with chlorophyll) obtain carbon from carbon dioxide. This results in the oxygen in the carbon dioxide being set free, and it is this oxygen which is constantly being made available for respiration by both plants and animals. Thus it is that plants which contain chlorophyll provide both food and oxygen for the remainder of the plant kingdom and for the whole of the animal kingdom. All life is dependent upon the green plant.

## Living Things Need to Grow

Living things are not the only ones which grow larger. An icicle, or a crystal of salt suspended in a strong salt solution, may increase in size. This, however, is the kind of growth which depends upon the addition of new layers to the outside. The living animal or plant, on the other hand, takes in foods and uses them to make quite different substances. From these, through the division of individual cells, the body of the living thing is built up from the inside.

This growth by cell division takes place in plants and animals alike. Plants may continue to grow throughout their lives. Certain animals may continue to grow throughout their lives also, but there is a tendency amongst some classes to attain full growth early in their adult lives. Animals also tend to be compact and restricted to a definite shape, whilst plants spread out and have less definite shape. During growth some animals shed their skins periodically.

The most massive and also the tallest of all living things are plants – the Californian Redwood trees. The 'General Sherman' tree in the Sequoia National Park is the most massive, and the tallest is the Howard Libbey tree of Humboldt County, estimated to be over 111 metres high. In the British Isles, the tallest trees are silver firs reaching heights up to 55 metres.

The largest of all animals are mammals – the blue whales (also known as sulphur bottom whales). They are known to have grown to over 30 metres, and are believed to be the largest of all animals which have ever inhabited our planet, exceeding in size the prehistoric dinosaurs.

The largest bird is the African ostrich (Struthio camelus). Male specimens have been known to reach a height of some 2.7 metres.

The insects with the longest bodies are the stick insects (*Phasmidae*) of tropical regions. Body length is reputed to be about 33 centimetres. Indian atlas moths have a wing span of some 30 centimetres, and a beetle (*Macrodontia cervicornis*) which does not exceed about 15 centimetres in length, has the bulkiest body.

## Living Things Need to Have Young

Any species in which reproduction ceased to be the fundamental aim would obviously be on the way to extinction. Methods of reproduction have improved with evolution, the highest evolved being capable of the most advanced methods.

In the plant kingdom the three main methods of reproduction are

*a* by part of the parent growing into a new individual
*b* by means of spores
*c* by means of seeds

The most advanced method for plants having young is by means of seeds, and most of the plants with roots, stems and leaves reproduce by this method.

In the animal kingdom the three main methods of reproduction are

*a* by part of the parent growing into a new individual
*b* by means of eggs
*c* by giving birth to young animals capable of free movement

The last method is the most advanced and ensures a better chance of survival. Mammals are a class of animals which (with two exceptions) reproduce by this method. Certain other animals, *e.g.* some species of insects, fish, amphibians and reptiles, retain their eggs until the young have hatched out, and this is sometimes referred to as 'live-bearing'. It is an unfortunate term, as it sometimes leads to the impression that eggs are not alive. Generally speaking, reproduction by means of eggs is the commonest method in the animal kingdom.

The illustrations for Lesson 2 in the Pupils' Book show:

1   the need for oxygen, represented by an astronaut on the moon wearing a spacesuit and a back pack. The lower part of the pack contains amongst other items, facilities for supplying oxygen, and the upper part of the pack contains an emergency 30 minute supply of oxygen. This system was used during the Apollo 11 landing on the moon in July 1969 and during the subsequent Apollo landings, culminating in the Apollo 17 landing in December 1972.

2   *a* an animal feeding on another living animal, represented by a young salmon (par) feeding on a smaller fish
    *b* an animal feeding on part of a living plant, represented by an emperor moth caterpillar feeding on a willow leaf

    *c* the roots of a dahlia, together with the swollen tubers for food storage

3  *a* growth in the animal kingdom, represented by a young piglet and an adult sow

    *b* growth in the plant kingdom, represented by a young tree and a large mature specimen

4  *a* having young in the animal kingdom, represented by eggs being laid amongst grass roots by a crane fly

    *b* having young in the plant kingdom, represented by plumed seeds of the willow herb being blown by the wind from the open fruit. Flowers and unopened fruits are also shown.

## CODE

1   Collect specimens of living things for the plant and animal sections of the science table.

2   Collect specimens of dead things for the plant and animal sections of the science table.

3   Observe any muscular movements which show that an animal is assisting the absorption of oxygen by breathing, *e.g.*

    *a* opening and closing of the mouth and gill covers of a fish

    *b* breathing of mammals such as humans, dog, cat

4   Observe anything connected with feeding, *e.g.*

    *a* different plant and animal foods on which various animals feed

    *b* the root 'hairs' through which food is absorbed into the roots of herbs, trees and shrubs

    *c* drooping in plants which lack water

5   Observe any signs of growth, *e.g.*

    *a* carrot tops in shallow water

    *b* pea or bean seeds planted on moist cotton wool or soil

6   Collect for future use any examples of

    *a* seeds

    *b* egg cases

### Written Work

1   Living things have four main <u>needs.</u>

2   They all need <u>oxygen</u> and food.

3   They need to <u>grow</u> and to have <u>young.</u>

4   Oxygen is one of the chief <u>gases</u> of the air.

5   From seeds grow <u>plants.</u> From eggs grow <u>animals.</u>

## 3  LIVING PLANTS ON LAND

### HERBS: TREES: SHRUBS: SIMPLE PLANTS

**Demonstration Material**

1  Any woody twigs from trees or shrubs.
2  Any terrestrial herb – either the whole plant or a section of the stem. An unwanted plant such as groundsel will serve.
3  Any terrestrial simple plants, such as
   *a* green algae, found on damp wood, brick or stone
   *b* common moss or lichen
   *c* fungi of any kind – toadstools, tree fungus, moulds or mildews.

**Sample Link Questions**

1  What is the difference between living animals and living plants? (*Living animals can move from place to place*)
2  What are the three homes of plants? (*Sea water, fresh water, land*)
3  What are the biggest living things in the world? (*Trees*)
4  What are the three main parts which most plants have? (*Roots, stems and leaves*)
5  What is the colour of most living plant leaves? (*Green*)
6  What happens to plants in the winter? (*Some die; some rest*)
7  What name do we give to trees which keep most of their leaves in winter? (*Evergreen*)
8  What name do we give to trees which have no green leaves in winter? (*Deciduous*)

**Relevant Information**

In Pupils' Book 1 the three homes of plants were said to be sea water, land and fresh water. Also in Pupils' Book 1 roots, stems and leaves were introduced as the three main parts found on most of the plants normally observed by children.

The main points of this lesson are that

1 there are three kinds of plants with roots, stems and leaves, namely *herbs, trees* and *shrubs*;

2 plants with no true roots, stems and leaves are referred to as *simple plants*;

3 examples of all these kinds are found on land.

## Plants in General

There are two kingdoms of living things – the animal kingdom and the plant kingdom. To children, the observable difference between non-microscopic forms of animal and plant life is that living animals are capable of movement from one place to another of their own choosing, whereas living plants are not. The term *plant* is frequently misused to segregate one group of plants from another, in such phrases as 'trees and plants', 'weeds and plants'. Further confusion has been caused by phrases such as 'fruit and vegetables', 'trees and flowers', whereas trees, weeds, fruits and flowers are of course all plants or plant parts, and the word 'vegetable' is – beyond the precincts of the greengrocer's shop – synonymous with 'plant'.

### *Plants with True Roots, Stems and Leaves – Herbs, Trees and Shrubs*

Most of the plants normally observed by children have true roots, stems and leaves. These are the characteristics of the land plant, for with them it can obtain the essential water from the soil and raise its leaves into sunlight and air.

A *Herbs.* These are not only a restricted group of plants used for medicinal purposes; the term has of course a wider meaning. A herb is any plant with true roots, stems and leaves, which does not possess the kind of woody stem found on trees and shrubs.

No part of a herb persists regularly above ground-level. Where winters are comparatively mild, some herbs *may* be found with parts above the ground, but many die off down to ground-level. *Annual* herbs die off completely after one season of growth. This is the case with many of the seed-bearing herbs, whose seeds remain dormant throughout the winter.

*Biennial* herbs live for two years. The seed puts out roots and grows during the first year, and then stores food to use the following year for flowers, fruit and seeds. Perennial herbs live for more than two years. In regions where the climate imposes seasonal growth, the living part of the herb may be entirely below ground during the winter – usually in some form of underground stem.

Herb-Peter is another name for the primrose, Herb-Trinity is the pansy, and Herb-Robert is a woodland geranium (*Geranium robertianum*).

B *Trees*. Trees have woody stems which persist above ground-level during the winter. Each tree has a single main stem, sometimes referred to as the trunk or bole. Trees are conspicuously land plants and are all perennials, living for more than two years.

C *Shrubs*. A shrub is like a tree, except that it has more than one main woody stem growing up from the roots. The term *bush* is often applied to a shrub with a dense growth of branching stems. Not all shrubs are bushy, and not all shrubs are big. The rose, privet, gorse, heather, mistletoe and ivy are typical shrubs.

Trees and shrubs generally live longer than herbs, and amongst them are to be found the oldest as well as the largest of all living things. A bristlecone pine growing in Eastern Nevada has been found to be about 4900 years old. Some of the Californian Redwoods are estimated to be between 3500 and 4000 years old with a possible life span of 6000 years. In the British Isles, the yew *Taxus baccata* is believed to be the plant with the longest life span. The oldest known, at Fortingall in Perthshire, is claimed to be over 3000 years old.

A plant which is a herb in certain conditions may not always grow as a herb in others. The castor bean plant, for example, grows as an annual herb in temperate regions, but in the tropics it may grow as a woody-stemmed perennial. Likewise, the hawthorn, willow and birch may grow in most areas as trees, but in colder northern regions they may be stunted into shrubs.

*Plants with No True Roots, Stems and Leaves – Simple Plants*

The four major classes are:

A *Algae*. These, like herbs, trees and shrubs, have chlorophyll, and therefore feed on inorganic (never-alive) foods. Some are one-celled, *e.g.* pleurococcus, which forms the green 'powder' to be found on damp walls and unprotected wooden fences; some form chains of cells, *e.g.* vaucheria, which is often found on the soil in plant pots; and others are more complicated, consisting of whole colonies of cells. All the seaweeds are algous plants.

B *Mosses and Liverworts.* These plants exhibit signs of the development of simple leaves and water-conducting stems and also rhizoids. (A rhizoid is a simple hair-like structure which has some of the functions of a root.) As mosses and liverworts possess chlorophyll, they therefore feed on inorganic foods as do the algae. Mosses especially are very common in damp places. They should not be confused with clubmosses, as the latter are herbs.

C *Fungi.* These have no chlorophyll, and therefore have to obtain their food partly made up, by feeding either on other plants or on animals. The *parasites* feed on living animals or living plants, and the *saprophytes* feed on dead animals or dead plants. They include the moulds, mildews, rusts, blights, mushrooms and toadstools.

A lichen is a combination of an alga and a fungus growing together.

D *The Bacteria.* These plants, which are responsible for what we call decay, are, like the fungi, lacking in chlorophyll. Some feed on living things and are therefore harmful. Certain of them are responsible for diseases, and tooth decay is an example of their activities. Most bacteria are useful however, and are responsible for the breaking down of dead animal and plant parts. Bacteria are not introduced until Book 4.

Although many simple plants are too small to be seen individually without magnification, they frequently form masses which are easily visible.

Illustrated in Lesson 3 in the Pupils' Book are the following land plants.

*Plants with Roots, Stems and Leaves*

*Herbs*

1   *Fern* – a representative of the coal-age plants. Ferns are chief amongst the spore-bearing herbs of today. There are over 6000 species in the world at present. Most of them develop their spores in spore cases on the underside of the leaves, which for most species, are compound.

2 *Wheat* belongs to the grass family. It is one of the oldest economic cereals in the world. There are many varieties. The clusters of flowers are wind-pollinated.

3 *Dahlia* – a perennial. The plants vary from 25 cm to 2 m in height. Compound flower heads of many colours are produced (c.f. composite heads, page 225).

## Trees

1 *Pine*. This represents the trees and shrubs which have young by means of cones and seeds. They often live in conditions which are too severe for other trees. Varieties are found in the sparse soil of mountain slopes as high as the timber line and also in the coldest zones in which trees grow. Pines are evergreen.

2 *Poplar*. This is representative of those trees and shrubs which have young by means of flowers, fruits and seeds. There are six species of poplars common in the British Isles, all deciduous. The one illustrated is the Lombardy poplar.

## Shrubs

1 *Rose*. The many varieties of garden rose shrubs are amongst the most popular of cultivated shrubs. They are deciduous.

2 *Privet*. The common European privet illustrated is used for hedges and ornamental purposes. It is a deciduous shrub, sometimes considered to be semi-evergreen, owing to its habit of retaining some of its old leaves until after the new ones have made their appearance in the spring.

3 *Heather*. Heaths and heathers are shrubs of the mountain and moorland. They are evergreens, common in the British Isles. There is no part of the year in which some variety of heather cannot be found with flowers.

## Simple Plants

### Algae

Algae are not predominantly land plants. The best-known are found in fresh or sea water, *e.g.* seaweed. Those found on land usually look like green coatings in soil, stones, old plant pots, etc.

## Mosses

Mosses do not have true roots, stems and leaves, although some of their thread-like growths may give the appearance of such. They have young by means of spores, and the spore-bearing capsules are shown. They are often found in carpet formation, consisting of thousands of plants. They are found in situations where they have access to moisture. Peat is the remains of dead moss.

## Fungi

The fungi shown are bracket fungi on a tree stump, toadstools, and mould. The main body of such plants consists of masses of fine filament-like threads (not roots) which are usually buried unseen in whatever host is serving as food. It is the spore-bearing reproductive structures which are raised into view, and which are therefore the only parts of the plants commonly seen.

## CODE

1  Collect specimens for the science table under general headings:
   a simple plants (no roots, stems or leaves)
   b herbs
   c trees and shrubs (mainly their twigs and leaves)
2  Observe how the woody stem of a tree or shrub differs from the soft stem of a herb.
3  Observe that a tree has one main woody stem, whereas a shrub has more than one main woody stem.
4  Observe moulds and mildews under a magnifying glass. Bread mould is easily cultivated by moistening a piece of bread and leaving it in a warm place for a few days. Without this encouragement, its growth takes a little longer.

## Written Work

1  Trees and shrubs have woody stems.
2  Trees have one main woody stem.
3  Shrubs have more than one main woody stem.
4  The stems of herbs are not woody.
5  Simple plants have no roots, stems or leaves.

# 4     PARTS OF HERBS, TREES AND SHRUBS

## FLOWERS, FRUITS AND SEEDS

### Demonstration Material

Any or all of the following:

1 Section of plant with flowers, fruits and seeds, *e.g.* snap-dragon, nasturtium, willow-herb.
2 *a* Flowers.
   *b* Fruits and seeds of various plants, *e.g.* horse chestnut, oak, pea, bean, poppy, buttercup, various berries.

### Sample Link Questions

1 What are the three kinds of plants with true roots, stems and leaves? (*Herbs, trees and shrubs*)
2 What do we call plants which have no true roots, stems and leaves? (*Simple plants*)
3 What is the main way in which trees and shrubs differ from herbs? (*Trees and shrubs have woody stems*)
4 What is the difference between a tree and a shrub? (*A tree has one main woody stem growing from the roots. A shrub has more than one*)
5 What do we call the main woody stem of a tree? (*Trunk*)
6 Do all plants have flowers? (*No*)
7 What are the four seasons of the year? (*Autumn, winter, spring, summer*)
8 What are the four main needs of living things? (*Oxygen, food, to grow, to have young*)
9 Are seeds and eggs alive or dead? (*Alive*)
10 What kind of living things grow from seeds? (*Plants*)

### Relevant Information

The main points of this lesson are:

1 most herbs, trees and shrubs can have flowers, fruits and seeds. Simple plants never can;

2   a flower grows from a bud on a stem;
3   a fruit is part of a flower which has grown;
4   seeds grow inside the fruit which serves to protect them.

The earliest plants were of course simple plants which lived in water but, following millions of years of evolution, some of these simple plants made their appearance on land. It is from these that the plants with true roots, stems and leaves evolved. Roots, stems and leaves were a necessary development of the land plants. Roots hold a plant firmly in the soil, and take in water and dissolved salts. Leaves manufacture starch, utilising carbon dioxide in the presence of sunlight. Stems raise the leaves into the air, and act as conducting agents. By the time the coal age had been reached, the dominant plants were those with true roots, stems and leaves, in the form of clubmosses, horsetails and ferns, reproducing by means of spores. However, the fundamental purpose for the existence of any living species is the propagation of its own kind, and indirectly from the ferns, evolved a more advanced group – the seed-bearing plants. They had their heyday during the period of the giant reptiles, and their surviving representatives are the cone-bearing trees and shrubs of today. Their seeds, however, are exposed naked on the scales of cones – a not too satisfactory state of affairs which led by evolution to the development of plants which had young by means of flowers, fruits and seeds, the latter being protected and completely enclosed by an ovary wall (the fruit).

Briefly then, the plants with roots, stems and leaves may be classified according to methods of reproduction as follows:

1   Those which have young by means of spores: clubmosses, horsetails and ferns. These are three classes of herbs.
2   Those which have young by means of cones and seeds: the class of coniferous trees and shrubs.
3   Those which have young by means of flowers, fruits and seeds. This class embraces most of the herbs, trees and shrubs.

The plants with flowers, fruits and seeds are the most highly evolved of the plants with roots, stems and leaves. The first probably appeared on the world about 125 million years ago, since when, they have advanced over a great part of the globe, sweeping to extinction many of their primitive relatives.

## Flowers

A flower is the reproductive structure of a plant which has young by means of flowers, fruits and seeds. Flowers are believed to have evolved from cones; in consequence, cones are often referred to as primitive flowers.

Although spring and summer are generally accepted as the seasons for flowers, flowers are available during all seasons. Various species of heather, for example, have flowers at different times during the year.

On some plants the flowers grow singly from the stem. On other plants they grow together in clusters. Lesson 25 in this book is about flowers which grow on their own and flowers which grow in clusters.

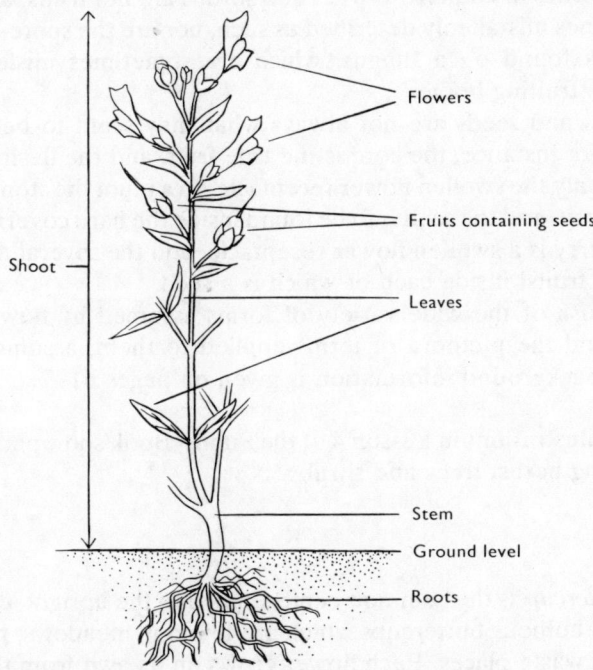

Some flower-bearing herbs live in fresh water, where their flowers either are raised on stems above the surface of the water or float on the surface itself. A few, *e.g.* hornwort and vallisneria, actually produce flowers below the surface. Flower-bearing plants have not established themselves in the sea, although a few species of herbs, such as the eel grasses, have managed to adapt themselves to conditions in salt marshes which are penetrated by tidal water.

## Fruits and Seeds

A true fruit is the ripened ovary of a flower. The wall of a fruit completely encloses and protects the seeds which are developed within it. Some fruits have a single seed; others have many seeds. On certain kinds of plants the ripened fruits open to allow their seeds to escape, but on the majority the fruit becomes detached from the plant, and the seeds are released afterwards. Not every flower produces a fruit – only those possessing pistils (female parts), the ovaries of which develop into fruits after fertilisation.

The cones of coniferous trees and shrubs are not fruits, although sometimes mistakenly described as such, nor are the spore-bearing surfaces found on a fungus, which are sometimes misleadingly termed 'fruiting bodies'.

Fruits and seeds are not always what they seem to be. In the apple, for instance, the core is the true fruit, and the fleshy edible part is only the swollen flower receptacle. In a plum the stone is part of the fruit, and the seed is to be found inside the hard covering. The strawberry is a swollen flower receptacle, and the several tiny pips are the fruits, inside each of which is a seed.

Because of the wide variety of forms assumed by flowers and fruits and the plethora of terms applied to them, a summary of useful background information is given on pages 61–7.

The illustrations in Lesson 4 of the Pupils' Book show parts of the following herbs, trees and shrubs.

## Herbs

1 *Buttercup* is the common name applied to the upright, creeping and bulbous buttercups which are found in meadows, pastures and waste places. Each flower grows on its own from the stem and has five sepals and five petals. A number of single-seeded fruits are found grouped together on one receptacle.

2 *Pea*. The flowers of this large family are irregular. They have a large standard petal at the back, two side wing petals, and two smaller petals which make a keel at the bottom. The fruit is a pod which splits when ripe, to allow the seeds to be scattered.

3  *Poppy*. The simple single brilliant red flowers are common in wheatfields in June and August. The large globular fruit is a capsule. It is through holes in the top of this that the seeds are scattered during windy weather, thus giving it the name of 'pepper pot'.

4  *Strawberry*. The red fleshy edible part – sometimes called a false fruit – is really the swollen flower receptacle. It is the tiny pips that are the true fruits. Each of these contains a single seed.

## Trees and Shrubs

1  *Apple* (a tree). The crab apple is the progenitor of the cultivated apple which customarily has pink and white flowers. The fleshy edible part is a false fruit – a swollen flower receptacle. The true fruit is the core with the seeds inside it.

2  *Rose* (a shrub). The rose illustrated is one of the many cultivated varieties grown for decorative purposes in British gardens and parks. The true fruits are found inside the swollen flower receptacle which turns red and is called a 'hip'. Each of the true fruits contains a single seed.

3  *Oak* (a tree). The flowers appear in clusters known as catkins in May. The fruit is the acorn, and the rough cup which holds it is a development from the sepals. There should, strictly speaking, be six seeds, although only one usually develops. Some acorns, however, have been known to contain two.

4  *Horse chestnut* (a tree). The flowers grow in clusters – sometimes called 'candles' – and appear in May. They may be white, white and pink, or pink. The fruits are thick, spherical and spiky, and contain reddish-brown seeds, commonly known to children as 'conkers'.

## Summary of Background Information Concerning Flowers, Fruits and Seeds

### The Main Flower Parts

A flower is a section of stem specially modified for the production of seeds. Its structure is closely related to pollination methods. The flower itself consists of a receptacle to which are attached all or some of the following four parts or floral organs: sepals, petals, stamens, pistils.

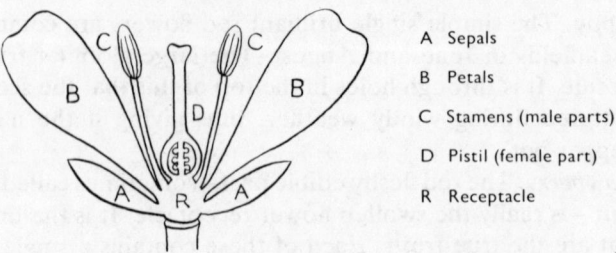

A  Sepals

B  Petals

C  Stamens (male parts)

D  Pistil (female part)

R  Receptacle

DIAGRAMMATIC STRUCTURE OF A FLOWER SHOWING ALL PARTS

*Sepals* are the outer leaf-like parts. They may be separate and free from one another or united along their edges. They are usually green but on some flowers they have other colours. Their main function is to protect the other floral parts before the bud opens. They sometimes fall off before the bud opens, as they do on the poppy, but on other plants they either remain until the petals fall or persist during the development of the fruit, as on the tomato and orange.

*Petals* form an inner ring of leaf-like parts. They too may be separate and free from one another or united along their edges. They may be white – especially on flowers which are pollinated by the wind – or of some bright colour which attracts insects, where these animals are responsible for pollination. In a collection of flowers from different plants the most noticeable differences will be those of shape, colour and arrangement of petals.

*Note*

The sepals and petals together are known as the *perianth*. The term is used, for example, to describe the collection of whorls on the bluebell, tulip and crocus, where the sepal and petals cannot be distinguished.

*Stamens* are found within the ring of petals. They may be attached to the receptacle or to other floral organs. A stamen is a male part, and on many plants each flower has a definite number of them.

*Pistils* are the innermost of the floral organs. They may be green or brightly coloured. A pistil is a female part, and on many plants each flower has only one. There are some plants, however, whose flowers possess more than one, *e.g.* buttercup. The pistil itself is composed of one or more carpels. Carpel is the term used to include an ovary (containing one or more ovules) together with the stigma, which is the receptive surface for pollen. It is the ovules which develop into seeds after fertilisation.

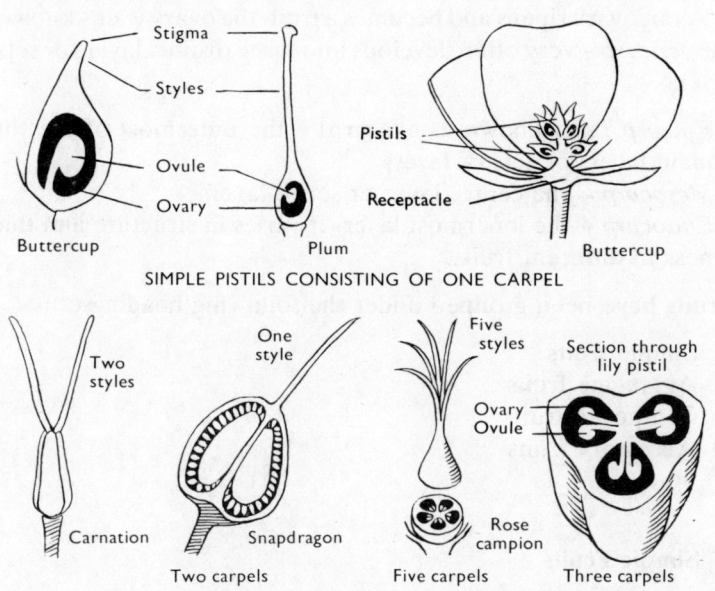

SIMPLE PISTILS CONSISTING OF ONE CARPEL

COMPOUND PISTILS CONSISTING OF SEVERAL JOINED CARPELS

If the pistil consists of one carpel, it is termed simple. If it consists of more than one carpel, then it is termed compound. When the pistil is compound, there may be several styles although the ovaries combine to make a compound ovary. A pea flower has a simple pistil consisting of one carpel. A buttercup flower has several simple pistils each consisting of one carpel. A tulip flower has a compound pistil consisting of three carpels fused together.

Stamens and petals drop off most flowers almost immediately after fertilisation, and the ovary begins to enlarge. Sepals may also drop off, but on some plants their withered remnants remain at one end of the fruit.

## Fruits and Seeds

After fertilisation the complete structure developed by the ovary is referred to as the *fruit*. Within the fruit the ovules develop into

63

seeds. A fruit therefore is a matured ovary, and a seed is a matured ovule, which develops inside the fruit. Some fruits contain one seed, and some contain more than one, depending upon the number of fertilised ovules.

As an ovary ripens and becomes a fruit, the ovary wall – known as the *pericarp* – very often develops into three distinct layers or sets of layers.

*a Epicarp* (also known as exocarp) – the outermost of the three main layers or sets of layers

*b Mesocarp* – the centre layer or set of layers

*c Endocarp* – the innermost layer. It varies in structure and thickness in different fruits.

Fruits have been grouped under the following headingse;

A  Simple fruits
B  Aggregate fruits
C  Composite fruits
D  Accessory fruits

## A  Simple Fruits

A simple fruit results from the ripening of one single ovary, and most fruits come into this category. The main types are:

### 1  *Fleshy Fruits*

In these the ovary wall (pericarp) becomes soft and fleshy. The eventual decomposition of this fleshy tissue allows the seeds to escape.

*a Berry*. In berries, the seeds are immersed in pulp or flesh. Examples are: orange, currant, cucumber, water melon, tomato, grape, grapefruit, lemon, gooseberry, pomegranate.

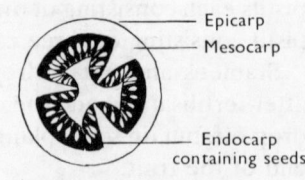

Epicarp
Mesocarp
Endocarp containing seeds

BERRY (Tomato)

*b Drupe*. In drupes, the endocarp becomes hard and is often known as a stone. Plums, cherries, apricots and peaches are drupes, and also such fruits as the coconut and almond. Some drupes have more than one stone; the holly 'berry' for example has four. A

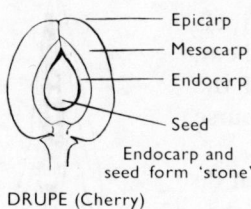

DRUPE (Cherry)

Epicarp
Mesocarp
Endocarp
Seed
Endocarp and
seed form 'stone'

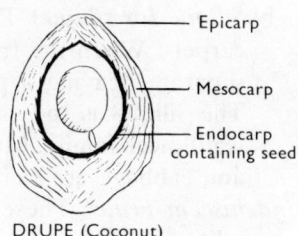

DRUPE (Coconut)

Epicarp
Mesocarp
Endocarp
containing seed

stone usually contains one seed, but in some two or three seeds may be found.

## 2 *Dry Fruits*

In these the pericarp becomes somewhat dry and may even be hard and brittle. They are of two kinds.

*a* Dehiscent fruits (which split open to allow the seeds to escape)
*b* Indehiscent fruits (which do not split open)

*Dehiscent Fruits*. At maturity, these fruits split open along definite seams, exposing several or many seeds.

    i *Legume or Pod*. This is formed from one carpel, and splits open along two seams. Examples are pea and bean.

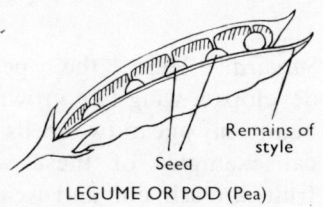

Remains of style
Seed
LEGUME OR POD (Pea)

FOLLICLE
(Larkspur)

    ii *Follicle*. This too is formed from one carpel, but it splits open along one seam, as in the columbine, peony, larkspur, marsh marigold, and monkshood.

    iii *Capsule*. This consists of two or more fused carpels, and it may split open in a variety of ways. Examples are poppy, lily, tulip, violet, antirrhinum, willow, poplar, iris, scarlet pimpernel and primrose.

5 carpels    3 carpels

CAPSULE (Pimpernel and violet or pansy)

iv *Siliqua (or silique)*. This consists of two fused carpels. When the fruit matures, the carpels separate, leaving a partition between them. The siliqua is the characteristic fruit of the wallflower family. Mustard, shepherd's purse and cabbage are examples

*Indehiscent Fruits*. These are dry fruits which do not split open along any definite seam when they are mature. They usually enclose just one or two seeds.

SILIQUA
(Wallflower)

i *Achene*. An achene has only one seed. Examples are the sunflower, dandelion and buttercup.

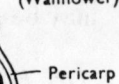

Pericarp

Seed

ACHENE (Buttercup)

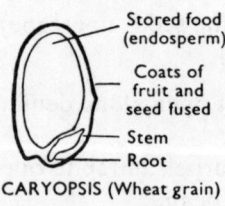

Stored food
(endosperm)

Coats of
fruit and
seed fused

Stem

Root

CARYOPSIS (Wheat grain)

ii *Caryopsis*. Sometimes known as grain, a caryopsis has only one seed which is fused to the inside of the pericarp, as in the grass plants, *e.g.* maize, wheat and oats.

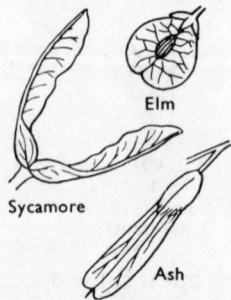

Elm

Sycamore

Ash

SAMARA (Winged fruit)

iii *Samara*. Here the pericarp develops a wing-like growth and bears only one or two seeds. Typical examples of these winged fruits are ash, elm and sycamore.

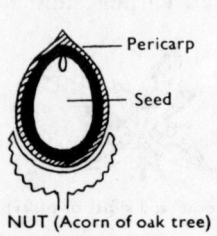

Pericarp

Seed

NUT (Acorn of oak tree)

iv *Nut*. Nuts are fruits containing one seed surrounded by a very hard woody pericarp. Examples are oak, hazel, beech and sweet chestnut. Some fruits and seeds referred to as nuts are not nuts in the botanical sense, *e.g.* peanut, walnut, coconut, horse chestnut.

## B  Aggregate Fruits

An aggregate fruit develops from a single flower with many separate pistils. A raspberry or a blackberry, for example, is a collection of drupes; a buttercup's fruit is a collection of achenes.

Buttercup

AGGREGATE FRUITS

Mesocarp
Seed

Blackberry

## C  Composite Fruits

This is a cluster of several fruits developed from several flowers which were themselves clustered together on the same inflorescence. Mulberries and pineapples are examples of composite fruits.

Mulberry

COMPOSITE FRUIT

## D  Accessory Fruits

These are not true fruits, but are fruits combined with other floral parts which are sometimes referred to as the 'fruit'. In the strawberry, the red fleshy part is a swollen receptacle, and the pips are individual achenes. A rose hip is a fleshy receptacle, and the fruits inside are achenes. In apples and pears the core is the true fruit containing the seeds, and the fleshy edible part is a swollen receptacle. Apples and pears are known as *pomes*.

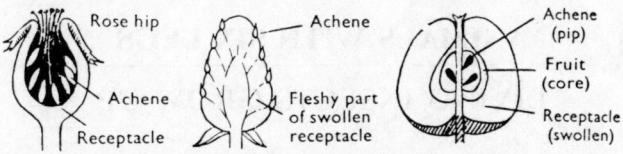

ACCESSORY FRUIT (Rose, strawberry and apple)

*Note*

A banana is a fruit in which the ovules have not developed into seeds. These are visible as small brown specks down the centre. The edible sweet chestnut is a fruit, the seed being the part which is eaten.

## CODE

1 Collect and observe autumn flowers, fruits and seeds.
2 Collect fruits and seeds under the general heading of 'Fruits and Seeds'.
3 Observe the true seeds inside an acorn or inside a plum, cherry or almond stone by removing the shell.
4 Observe that, when stored unpreserved, flowers and fruits eventually die (in some cases attacked by a fungus), whereas the seeds may remain unspoilt.

*Note*

Seeds which are likely to be used in future lessons (*e.g.* peas and beans) should be stored in a container which allows them fresh air. Fixing to the container a label bearing the date, serves as an indication of their length of life. Dried pea and bean seeds may, of course, be stored for years awaiting germination when required.

## Written Work

1 Most herbs, <u>trees</u> and shrubs have flowers.
2 A flower grows from a <u>bud</u> on the stem.
3 A fruit grows out of a <u>flower</u>.
4 <u>Seeds</u> grow inside the fruit.
5 <u>Simple</u> plants do not have flowers, fruit and seeds.

# ANIMALS WITH SIX LEGS

## LIVING INSECTS GROW UP

### Demonstration Material

1 Any living, mounted or preserved examples of the egg, larva and adult of any insect with a three-stage life, *e.g.* stick insect, earwig, dragonfly, mayfly, grasshopper, cricket, cockroach, aquatic bug.

2    Any living, mounted or preserved examples of the egg, larva, pupa and adult of any insect with a four-stage life, *e.g.* moth or butterfly, beetle, two-winged fly, caddis fly, ant, bee, wasp.

## Sample Link Questions

1    What are the four main needs of living things? (*Oxygen, food, to grow, to have young*)

2    What are the two main things which living animals and plants need in order to grow? (*Oxygen and food*)

3    Do all living things grow bigger throughout their lives? (*No; some stop growing after a time*)

4    Are seeds and eggs alive or dead? (*Alive*)

5    Which grows from an egg – a living plant or a living animal? (*A living animal*)

6    What is a difference between living animals and living plants? (*Living animals move from place to place*)

7    What are animals with six legs called? (*Insects*)

8    How many wings do *most* adult insects have? (*Four*) Some have two; some have none.

9    How many feelers has an insect? (*Two*)

## Relevant Information

### Animals in General

Probably the principal distinction which can be drawn between animals and plants is that no animal is capable of converting inorganic material into organic material during feeding, and must therefore rely on food partly made up by plants. Directly or indirectly, therefore, the animal kingdom is dependent upon the plant kingdom. Other general differences can be noted however. For example, most animals move from place to place to find their food. (Species such as the acorn barnacle, sponge and sea-anemone which settle down to a sedentary life are exceptional.) Most plants on the other hand remain in the same place, with only exceptional species such as the microscopic *Chlamydomonas* being capable of self-propulsion. Again, animals tend to be compact and restricted to a definite shape, whilst plants spread out and have no definite shape. Most animals reach a state of full growth, but most plants continue to grow throughout their lives.

The members of the animal kingdom are grouped according to certain characteristics into classes. The animal classes of mammals, birds, fish and insects were the subject of lessons in Pupils' Book 1. Amphibians, reptiles, spiders, crustaceans and the many-legged centipedes and millepedes are the subject of separate lessons in this book.

*Mammals* – the class of animals to which humans belong – are the only animals whose young need to be fed on milk, and they all have hair. Where there is a lot of hair, it is termed *fur*. Whales are mammals which have returned to feed in the sea, and bats are the only mammals which can fly.

*Birds* are the only living things to bear feathers. Their legs still bear scales like the reptiles from which they are descended, and their front limbs are developed as wings.

*Fish* are animals with gills and paired fins. They were the first animals to have backbones. Most kinds, but not all, have scales.

*Amphibians* – the first of the back-boned animals to walk on land – in general live part of their lives in water and part on land. They include frogs, toads and newts.

*Reptiles* are animals with scaly skins and with lungs. The prehistoric monsters were reptiles, and present-day representatives include the crocodiles, turtles, snakes and lizards.

Fish, amphibians, reptiles, mammals and birds are all *vertebrates, i.e.* animals with an internal skeleton. Mammals and birds evolved from primitive reptiles; reptiles evolved from primitive amphibians; amphibians evolved from primitive fish. The ancestors of fish are presumed to have been related to the ancestors of worms.

*Insects* are animals which have six legs in the adult stage. Most adults have four wings, some have two, and certain species have none.

*Spiders and harvestmen* are animals with eight legs. They belong to the *arachnid* class, which also includes scorpions and mites.

*Crustaceans* are animals with crusty skins. They are represented in the sea by the crab, lobster, prawn and shrimp; in fresh water by the crayfish, freshwater shrimp, water louse and pond flea; and on land by the woodlouse.

*Centipedes and millepedes* are land animals with many pairs of legs all the way along the body. Centipedes have one pair per segment, and millipedes have two pairs to nearly every segment.

Insects, arachnids, crustaceans, centipedes and millepedes have jointed limbs, but no internal skeleton. Instead, they have a tough outer skin (*the exoskeleton*) made of a horny substance called chitin, which protects the soft inner parts. These four major classes of animals (often termed collectively *arthropods*) have their bodies divided into segments, and are believed to have evolved from the ringed worms.

## Simple Animals

The main groups of animals which appeared earlier in the evolutionary story than those mentioned above were relatively simple animals. As well as lacking other advanced features, they had neither an internal nor an external skeleton, and none had jointed limbs. Four of these groups of simple animals have prominent representatives in modern times. These are:

*Worms,* (animals with long soft bodies), which include a vast assembly of elongated, crawling, simple animals with representatives in the sea (*e.g.* lugworm and tube-building worms); in fresh water (*e.g.* leeches and flat worms), and on land (*e.g.* earthworm). A considerable range of species are parasites, *e.g.* tapeworms. Many animals which are not worms at all, are confusingly termed worms, *e.g.* slow-worm, glow-worm, mealworm, silkworm, woodworm and shipworm: ringworm – a disease of the skin – is caused by a fungus.

*Molluscs* are soft-bodied simple animals with representatives in the sea, in fresh water and on land. They include the slugs, snails, limpets, cockles, mussels, oysters, squids and octopuses, and presumably they evolved from the worms. The most outstanding feature common to most of them is the hard, limestone shell.

Mouth and stomach animals (*Coelenterata*) which include the jellyfishes, the sea anemones and corals, freshwater hydra, Portuguese

71

man-o'-war, sea-fir and various small animals known as sea-gooseberries or comb-bearers. The body of such animals is basically a simple sac with a single opening at one end through which food is taken in and waste expelled.

Animals with spiny skins (*Echinoderma*) which are more advanced than the coelenterates, and include the starfishes, brittle-stars, sea-urchins, sea-cucumbers and feather-stars (also known as sea-lilies or crinoids). The skin is strengthened by small rods and plates of calcium carbonate (limestone) with an additional protection of spines, which give rise to the terms 'spiny-skinned' and 'hedgehog-skinned'.

Worms and Molluscs are the subject of separate lessons in Book 3.
   Mouth and stomach animals and animals with spiny skins are the subject of the final lesson in Book 4.


**Insects**

These are the most numerous of all animals and are found in every part of the world where terrestrial life is possible. It has been estimated that the total bulk of insects in the world is far greater than that of all other animal life put together. There are more than 20 000 species of insects to be found in the British Isles alone. Salt water does not seem to suit them, but there are many species in fresh water. No matter how tiny an insect is, it has a nervous system, a digestive system, a heart and blood, but no lungs. The word insect is derived from the Latin *insectum* meaning 'cut into', because these animals are divided into the three parts – head, thorax and abdomen. The most easily observable characteristic which distinguishes adult insects from all other animals, is that they all have six legs.

*Respiration.* Most insects respire by means of air holes (spiracles) in the side of the body which give access to hundreds of branching tubes extending to all parts of the body, even to the tips of the wings and antennae. Some aquatic insects respire by means of tracheal gills in the early stages of their life, but even these are compelled to come to the surface to obtain oxygen from the air once they reach the adult stage.

*Food.* Insects feed on a wide variety of animal and plant foods, and have very complicated mouth parts. There are two types of mouth:

*a* sucking mouth parts as on true flies, butterflies, moths, and bugs
*b* biting or chewing mouth parts as on beetles, dragonflies, crickets, grasshoppers, cockroaches and earwigs.

*Growth.* The growth from egg to imago (adult) follows one of two main courses:

A  Egg, larva, adult (the term *nymph* is sometimes used instead of the term larva).
B  Egg, larva, pupa, adult.

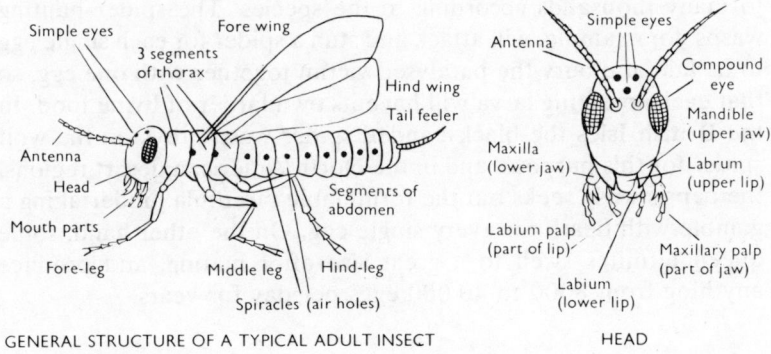

GENERAL STRUCTURE OF A TYPICAL ADULT INSECT       HEAD

## A *Insects with a Three-Stage Life (Egg, Larva, Adult)*

1  A few small, insignificant, and apparently primitive insects, *e.g.* the bristle-tails, silver fish and spring-tails, hatch into miniature replicas of the adults. These grow into adults which, being primitive, are completely wingless.
2  The majority of insects with a three-stage life hatch out into larvae which will have wings in the adult stage. This is sometimes referred to as an incomplete metamorphosis, and the larva is often referred to as a *nymph*. In some species a mode of life has been adopted which has made it unnecessary for the females to use wings, and therefore wings are developed only on the male. The stick insect with its near-perfect camouflage is an example.

73

## B *Insects with a Four-Stage Life (Egg, Larva, Pupa, Adult)*

Most insects pass through this kind of life, which is sometimes referred to as a complete metamorphosis. It is the development assumed by the most highly specialised insects and often involves several larval moults and several larval phases.

### Eggs

Whereas most adult female insects lay their eggs close to, on or even inside the food upon which the larvae will feed, the majority exhibit no further maternal instincts. Earwigs, ants, bees and wasps are, of course, the best-known exceptions to this generalisation.

The number of eggs deposited in one place may number from one to many thousands according to the species. The spider-hunting wasps, for example, will attack and stun a spider for each single egg to be laid and bury the paralysed victim together with one egg, so that each emerging larva will have its own larder of living food. In the British Isles the black-banded spider wasp preys on the wolf spider for this purpose, and in the North American desert regions, the pepsis wasp seeks out the formidable tarantula, undertaking a gamble with death for every single egg. On the other hand, some queen termites swell to a great size after mating, and produce anything from 8000 to 40 000 eggs per day for years.

### Larvae

Only in the larval stage does an insect grow. An insect with a three-stage life hatches out of the egg looking very much like a miniature adult, except for the lack of wings. The larva or nymph has mouth parts like the adult and usually has compound eyes. In insects with a four-stage life the larvae present quite a different appearance from the adults and hatch out into soft-bodied animals more like the worms from which they are descended. The change from a soft body to an armoured insect takes place in the pupal stage. In insects with a four-stage life the mouth parts in larva and adult may be entirely different, and in the larval stage the compound eyes are usually lacking.

The wings of an insect with a three-stage life develop under the skin of the growing larva. The larva sheds or moults its skin several times during growth and, following the final moult, the wings

# ANIMALS WITH SIX LEGS

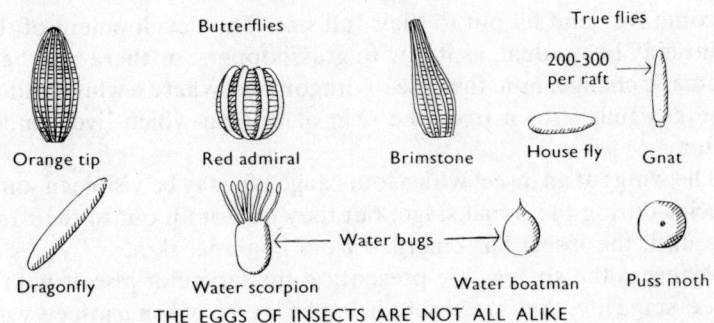

Butterflies

Orange tip    Red admiral    Brimstone

True flies

200-300 per raft

House fly    Gnat

Dragonfly    Water scorpion    ←— Water bugs —→    Water boatman    Puss moth

**THE EGGS OF INSECTS ARE NOT ALIKE**

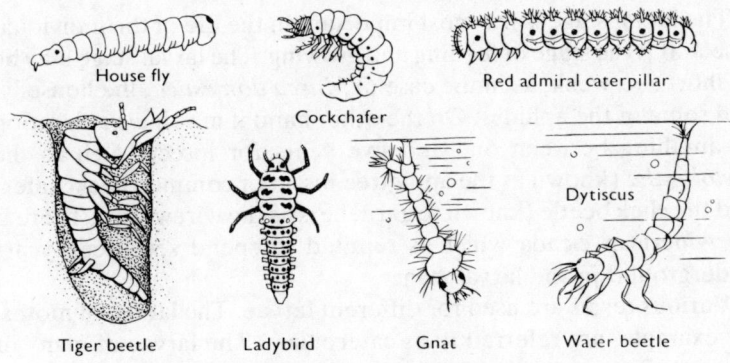

House fly    Cockchafer    Red admiral caterpillar

Dytiscus

Tiger beetle    Ladybird    Gnat    Water beetle

**THE LARVAE OF INSECTS ARE NOT ALIKE**

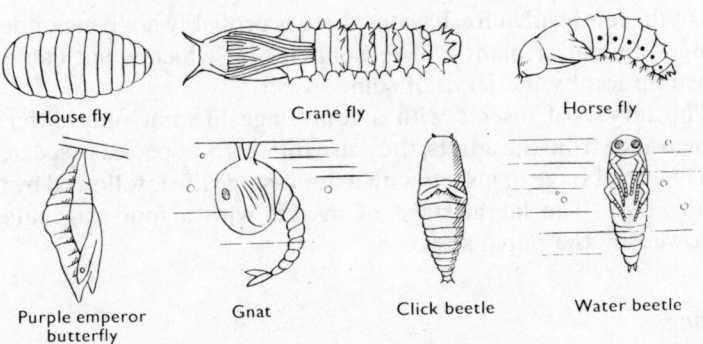

House fly    Crane fly    Horse fly

Purple emperor butterfly    Gnat    Click beetle    Water beetle

**THE PUPAE OF INSECTS ARE NOT ALIKE**

become free and fill out to their full size. The development of the adult may be gradual, as it is with grasshoppers, or there may be a dramatic change, as in the case of dragonflies, where a winged adult emerges fully grown from the skin of a larva which lived under water.

The wings of an insect with a four-stage life may be visible in some species during the pupal stage, but they do not fill out to their full size until the insect has emerged from its pupal skin.

Whereas the six legs are present on the larvae of insects with a three-stage life, they are absent from the larvae of some insects with a four-stage life. Moths and certain beetles, for example, have their six legs present in the larval stage. On the other hand the two-winged flies (*Diptera*), *e.g.* bluebottle and housefly, are legless in the larval stage.

The larval stage is the most important in the life of the individual insect. It is the stage of feeding and growing. The larval stage may be as short as a week, as in the case of *Musca domestica*, the housefly, and some of the aphides. On the other hand it may have a duration of anything between one and five years for insects such as the *Melolontha* (known as the appletree insect or common cockchafer) and the click beetle (known to gardeners as the wireworm). There is an American cicada which is reputed to spend seventeen years underground in the larval stage.

Various terms are used for different larvae. The larvae of moths, for example, are referred to as caterpillars. The larvae of many of the true flies are called maggots or gentles, and others, such as beetle larvae, are referred to as grubs and sometimes, confusingly, as worms, *e.g.* wireworm, woodworm, glow-worm. Similarly the larva of the silk moth is referred to as the silk worm.

On the land and in fresh water there is probably no living or dead thing – animal or plant – of reasonable size which is not liable to consumption by the larva of some insect.

The larvae of insects with a four-stage life not only differ in appearance from the adults, they also differ from species to species.

The larval stage of insects with a three-stage life is followed by the adult stage. The larval stage of insects with a four-stage life is followed by the pupal stage.


## Pupae

Where there is a pupal stage in an insect's life, it is one of reorganisation and often – but not always – of rest. The pupae of moths and butterflies, bluebottles and houseflies, beetles and ants remain dormant while the adult develops within the pupal skin. The pupae of gnats and midges on the other hand are free – swimming in their aquatic surroundings. The larvae of moths and butterflies spin a covering of silk for protection in the pupal stage. The last larval skin of the bluebottle and housefly thickens into a hard oval case, whereas beetle pupae remain soft with the limbs free of the body, and relying on their being hidden in the ground for protection. *Pupa* is Latin for *doll*, but not all pupae are doll-like.

## Adult

The adult stage (imago) is essentially one of reproduction. Once the propagation of the species is assured, the adults – especially the males – may die. An adult insect usually has a comparatively short life, sometimes not more than a few days. Mayflies, for example, may spend up to three years in reaching the adult stage and then be on the wing for perhaps less than twenty-four hours. Queen bees and worker ants may live for two or three years as adults, and it is claimed that certain queen ants have lived for up to fifteen years. The less fortunate male insects may die after a single mating; the female praying mantis has been known to consume the head of her spouse during the action of mating – with apparent indifference on both sides. When an adult insect feeds, it is not for the purpose of further growth, but only to supply energy necessary for movement and mating, material for eggs, and to replace liquid lost through evaporation. Some adult insects may continue to feed and survive for long periods, but the mayflies, and certain moths and caddis flies appear, as adults, to abstain from food altogether.

A wide range of sizes is to be found in the insect kingdom. Some tropical stick insects have a body length of about 33 centimetres, and the Indian atlas moth has a wingspan of about 30 centimetres. There are beetles almost 15 centimetres in length, and there are beetles tiny enough to crawl through the eye of a needle.

Insects with a three-stage life include the earwigs, the mayflies, the bugs, the dragonflies, and the grasshoppers, crickets and cockroaches. Insects with a four-stage life include the true flies, the beetles, the caddis flies, the moths and butterflies, and the ants, bees and wasps.

77

The illustrations for Lesson 5 of the Pupils' Book show:

1    Insects with a three-stage life
     *a* Common earwig: egg, larva, and male adult
     *b* Stick insect: egg, larva, and female adult
2    Insects with a four-stage life
     *a* Ladybird: egg, larva, pupa, and adult (seven-spotted scarlet variety)
     *b* Common wood ant, or red ant: egg, larva, pupa in cocoon, and a female adult without wings

### Earwigs

Up to 60 eggs are laid in holes in the ground during the early spring. The female guards them and may also protect the newly-hatched larvae. The larvae reach the adult stage by midsummer. The forewings of the adult are reduced to short protective wing cases, and the tail pincers are used to fold the delicate hind flight wings beneath these wing cases. The pincers are used to adopt an aggressive attitude. They also serve to distinguish the sexes, those of the male being longer and more sharply curved. The adults rest for the winter in holes – NOT in the ears of sleeping humans.

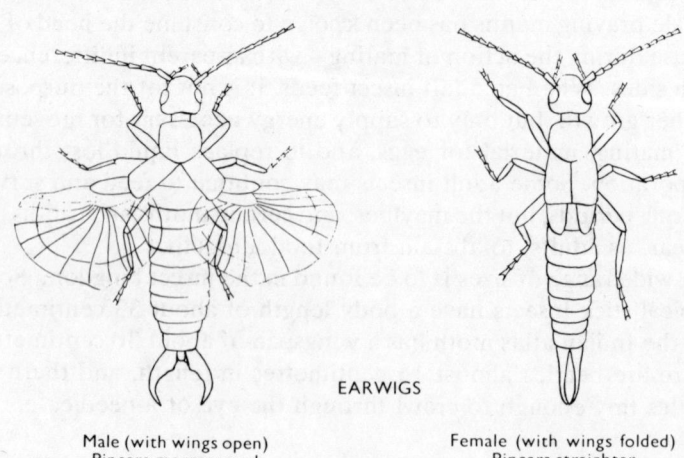

EARWIGS

Male (with wings open)          Female (with wings folded)
Pincers more curved                Pincers straighter

### Stick Insects

Up to 700 eggs are laid, according to species, and scattered at the base of the food plant. Each is a hard dark-brown capsule with a little green lid through which the larva eventually emerges. The larva looks like the adult, apart from size, and moults about six times during growth to the adult stage. Adult length varies from 12 to 23 centimetres, according to species, although some tropical insects attain 33 centimetres – being the longest insects in the world. The females are wingless and sluggish in movement. Males are more active and have wings, but are very rare and probably useless, as the females are able to reproduce without their eggs being fertilised by a male. In their imitation of plant stems, stick insects are an example of perfect camouflage in the insect world. Although they are all tropical or subtropical insects, they may easily be kept indoors in Britain, with privet or ivy for food.

## Ladybirds

Ladybirds belong to the beetle family of insects. Eggs are laid in small groups on aphid-infested plants, and the larvae feed for about three weeks on the aphids before pupating. The pupae are leathery and attached by their tails to leaves. The adult continues to feed on aphids. Some hibernate for the winter, often in groups under the bark of a tree or beneath wooden window sills. Different species may be distinguished by colouring and by the number and position of spots.

## Ants

Ants have reached the highest level of development amongst the animals without backbones. Ant eggs are very tiny. A female (queen ant), starting a new nest, lays them in an underground chamber and guards them. The larvae are eyeless, legless and helpless, and are fed by the mother until they pupate. The larvae of a number of species – including the wood ant – form cocoons in which to pupate. It is the pupal cocoons of the wood ant which are sold as fish food under the false title of 'ants' eggs'. The first adults to emerge are wingless females. These become the worker ants, excavating fresh galleries and attending to subsequent larvae and pupae, whilst the queen confines herself to the task of egg-laying. Subsequently eggs are laid from which hatch larvae destined to become winged females and winged males. After these have mated, the males die, and the females return to the home nest to produce further eggs, or begin new colonies of their own. In the adult stage, ants differ from other insects in having more than one waist. They are rarely found with wings. The worker ants develop without them; the males, which have four, die shortly after mating, and the queens, which also have four wings, discard them after mating, as they are no longer required.

*Note*

*Larva* and *pupa* are Latin terms. The plurals are *larvae* and *pupae*, and the adjectives are *larval* and *pupal*. In the Pupils' Book at this stage, only the singular nouns have been used, the use of the plural nouns and the adjectives being left to the discretion of the teacher.


## CODE

1    Observe any living, mounted or preserved specimens of:
      *a* egg, larva and adult of insects with a three-stage life;
      *b* egg, larva, pupa and adult of insects with a four-stage life.
2    Observe:
      *a* that the larva of an insect with a three-stage life is similar to the adult and that it has six legs;
      *b* that the larva of an insect with a four-stage life is quite different in appearance from the adult and may, or may not, have six true legs.
3    Observe any pupae which may be found.
4    Observe that it is only in the adult stage that an insect has wings.
5    Observe any adult earwigs found during the day under stones, and sometimes indoors. If it is desired to keep them, an insect cage containing an unimportant flowering plant may serve as a temporary home.
6    Observe how the larva of a stick insect (being an insect with a three-stage life), is similar to the adult, by rearing in an insect cage. Stick insects are amongst the most agreeable insects to keep in captivity. Reared in this country, they may be fed exclusively on a diet of privet leaves, but may also feed on ivy leaves. The hard brown eggs with their little green caps may be collected from the bottom of the insect cage, and stored for about five months when they will begin to hatch. They may be stored in small cardboard boxes or in tins with two or three small ventilation holes in the lid. Careless handling may be avoided by brushing the eggs on to paper with a soft paint brush.

7 Observe how the larva of a bluebottle (being an insect with a four-stage life), differs from the adult. Bluebottle larvae – maggots – may be bought for a few pence from shops supplying angling tackle. Mixed with a little sawdust, and kept in a ventilated jar in a reasonably warm room (classroom temperature is suitable), they will pass from the larval, through the pupal, and to the adult stage, without further attention.

8 Observe how an adult ladybird unfolds its two flying wings from under its two wing cases. Although children are very fond of ladybirds, they are difficult to keep for more than a short time in captivity unless a constant supply of aphids is available. It is best to release specimens after observation.

9 Preserve any required specimens in 5% formaldehyde in the usual way.

## Written Work

1 Eggs are laid by the <u>female</u> adult.
2 Only the <u>larva</u> of an insect grows.
3 In the stage of <u>pupa</u>, most insects rest.
4 Only <u>adult</u> insects can have wings.

**6** **TAKE CARE OF LIVING THINGS**

## LET INSECTS HAVE OXYGEN AND FOOD

### Demonstration Material

1 Large tin lid.
2 Large jar, *e.g.* toffee jar.
3 Ink bottle containing water.
4 Scissors, preferably strong and with a point.
5 Stem or branch stem with leaves, together with any insects which feed on those leaves, *e.g.*
    *a* privet and stick insects.
or *b* caterpillar and section from appropriate food plant.

6   Any made-up insect cages as illustrated in the text.

## Sample Link Questions

1   What are the three stages in the life of an insect with a three-stage life? (*Egg, larva, adult*)
2   What are the four stages in the lives of most insects? (*Egg, larva, pupa, adult*)
3   What does a larva need to do when it hatches out of its egg? (*Feed*)
4   Which is it that looks like the adult – the larva of an insect with a three-stage life, or the larva of an insect with a four-stage life? (*The larva of an insect with a three-stage life*)
5   At which stage does an insect grow bigger? (*Larva*)
6   Does an insect feed when it is a pupa? (*No*)
7   What do most insects do at the pupal stage? (*Rest*)
8   Which is the only stage at which an insect may fly? (*Adult stage*)
9   What are the four main needs of living animals and living plants? (*Oxygen, food, to grow, to have young*)
10  Which of these do living things need in order to grow? (*Oxygen and food*)

## Relevant Information

There was a lesson in Book 1 on Taking Care of Fish. The main purpose of this lesson is to show how to take care of captive insects, *i.e.* by catering properly for their oxygen and food requirements.

Oxygen and food are the two basic needs upon which growth and having young depend. When any living thing – animal or plant – is confined for observation purposes at a suitable temperature, and in suitable surroundings, it follows that oxygen and food are the two most important requirements for which provision must be made.

Animals and plants which belong to the land have no difficulty in obtaining oxygen, if there is sufficient ventilation to allow them a constant supply of fresh air. The chief problem which remains with terrestrial animals and plants is to provide them with a sufficiency of the right kind of food. Where animals which feed on living foods are kept, it is therefore important to keep the food alive until it is consumed.

This maxim of providing the right kind of food applies just as much to insects as it does to other animals. Young children, however, are not always so well-informed of this as they could be. This is why matchboxes, jam jars and dying leaves are all too frequently the tragic fate of collected caterpillars. This too is why cabbage leaves, for example, are sometimes fed indiscriminately to caterpillars, with little or no consideration for whether cabbage leaves are part of their normal diet.

There are insects which feed on living things and insects which feed on dead things. Those which feed on dead things are perhaps the easiest to provide for in captivity, and especially those which feed on manufactured foods containing dead material, *e.g.* biscuits, Bemax.

Of those insects which feed on living foods, some feed on plants or plant parts, and some on other animals. Obviously it would be ideal if the whole living organism could be provided for captive insects which feed on living things, but this is seldom practicable. It is not always convenient, for example, to supply the whole plant for insects which feed on a particular herb, tree or shrub. On the other hand, if leaves are supplied on their own, they will die quite rapidly.

The most suitable arrangement when keeping insects which feed on living herbs, trees or shrubs is to obtain as much of the plant as possible. Cuttings of the stem or branch stem bearing the leaves should suffice. If these stems are placed in water, they will feed the

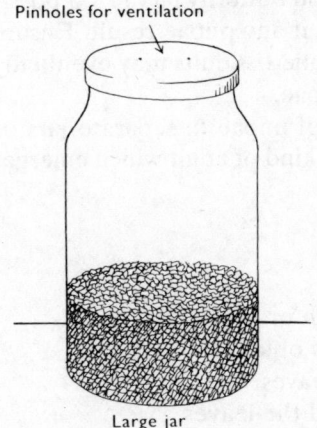

Pinholes for ventilation

Large jar

A type of home suitable for insects which do not feed on living foods, in this case flour weevils. Successive generations may be reared for years without any attention other than the occasional addition of oat-meal or Bemax.

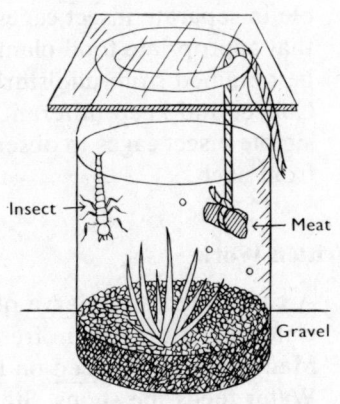

Insect

Meat

Gravel

A home suitable for aquatic insects, in this case the larva of a water beetle (Dytiscus). The meat on the string should be left in for a few hours only. The cover should be loose, and the bottom of the jar filled with gravel.

leaves, which in turn will ensure a fresh supply of living food for the captive insects. Twigs from trees and shrubs are particularly amenable in this respect.

Moth and butterfly larvae – caterpillars – have always fascinated children, and may easily be reared to the adult stage if suitably catered for. It is reasonably safe to assume that when caterpillars are found in numbers on a particular plant, then that species of plant is the right food for that caterpillar. Therefore if some of those caterpillars are to be kept, then it is that species of plant which must be provided for food.

It is obviously desirable to provide surroundings in which insects are easily visible. Commercially-made insect cages are often expensive. The 'cages' on page 83 (and those shown in the Pupils' Book) have proved cheap, effective and easy to make. They also serve as models for children who wish to keep insects successfully in their own homes.

## CODE

1 Demonstrate the setting up of a simple insect cage consisting of a jar, tin lid, bottle with water, and food plant, as shown on the pupils' page.
2 Observe different kinds of moth and butterfly larvae – if possible in separate insect cages to see if any pupae result. Ensure that appropriate food plant is supplied. Adults may eventually be obtained from undisturbed pupae.
3 Collect and keep different kinds of pupae in separate jars or simple insect cages to observe the kind of adult which emerges from each.

## Written Work

1 A caterpillar is the larva of a moth or butterfly.
2 Caterpillars breathe in fresh air to obtain their oxygen.
3 Many caterpillars feed on living leaves.
4 Water feeds the stems. Stems feed the leaves.

# 7    LIVING ANIMALS CAN MOVE FROM PLACE TO PLACE

## MOVING IN WATER, ON LAND AND IN THE AIR

### Demonstration Material

1 Any living specimens of animals – especially insects, birds, fish or mammals.
2 Any mounted or preserved specimens on which limbs, *e.g.* legs or wings, can be observed.

### Sample Link Questions

1 What is one difference between living animals and plants? (*Living animals can move from place to place*)
2 Which class of animals always feed their babies on milk? (*Mammals*)
3 How many legs have most mammals? (*Four*)
4 What are animals with feathers called? (*Birds*)
5 How many wings has every bird? (*Two*)
6 Which animals always have six legs when they are adults? (*Insects*)
7 What else do most adult insects have to help them to move from place to place? (*Wings*)
8 What do we call an animal which has both fins and gills? (*A fish*)
9 Which liquid do all living animals and plants need for food? (*Water*)

### Relevant Information

Points already taught in Pupils' Book 1 are:

1 the two kinds of living things are animals and plants;
2 those which are observed to move from one place to another are animals.

The main points of this lesson are:

1 living animals need to move from place to place for two main reasons:
 *a* to find food
 *b* to escape from enemies (which would use them for food)
2 various means have been developed for enabling animals to force themselves to move, in water, on land, and in the air.

The fundamental aim of any living species is the propagation of its own kind. The movement of living animals from one place to another is linked, directly or indirectly, with this aim. Movement results from

1 the need for food which will enable members of the species to grow to maturity and propagate their own kind;
2 the need for preservation, which will enable members of the species to survive, thus ensuring the propagation of their kind. In this respect the ability to move is used
 *a* to enable the individual to escape from danger – principally from some other animal seeking it for food;
 *b* to find suitable conditions for reproduction – where food is plentiful and where danger from other animals is reduced.

In order that they may propagate, living animals and plants must first grow to maturity. Growth to maturity depends upon the living organism obtaining oxygen and food. Obtaining oxygen gas seldom offers problems. To terrestrial animals and plants it is available as a gas mixed in the air. To aquatic animals and plants it is available as a gas dissolved in the water.

But although the energy requirements of plants and animals are both satisfied by the same gas – oxygen – the food requirements of plants and animals are not satisfied by the same foods. Only the green plants (*i.e.* those with chlorophyll) can exist entirely on never-alive foods, and when such plants are once established where there is a sufficiency of water and dissolved salts, they are not faced with the necessity of having to seek them in new locations.

However, although the green plants feed on inorganic foods, other living things rely for at least part of their food on either living or dead animals or plants. These consist of the non-green plants (bacteria and fungi) and the animals. Obviously they will starve once they have consumed the available material around them, unless:

1   more suitable food is carried to them
*or* 2   they are able to transport themselves to fresh supplies of
    food.

Once the food of a fungus, for example, is exhausted, it dies, and it is left to its spores – transported by moving animals, water or wind – to find more food, and thus to continue the species. Animals, however, have evolved means of transporting themselves from one food source to another. The means are more purposeful, and, where reproduction is concerned, less wasteful. Thus it is that the observable difference between non-microscopic forms of animal and plant life is that living animals are capable of movement from one place to another of their own choosing, whereas plants are not.

The search for food takes up a great deal of the time in the lives of most wild animals. Some animals feed on plants; others feed on the animals which feed on plants; and, of course, some animals feed on both animal and plant foods.

Some animals search for their food singly, others travel in companies. The movement of herds of herbivorous mammals such as reindeer and elephants is from one food grazing area to another. Many fish travel in shoals. Flocks of birds such as finches seek out seeds such as grain, and other foods. Swarms of locusts numbering millions at a time fly from one area of vegetation to another, denuding the land as they go.

Just as plant-eating animals move from one source of food to another, so carnivorous animals which feed on *them* need to move in order to catch them. Frequently they need to move faster.

When circumstances are such that successive generations of a species of animal are born where food is abundant, or have their food brought to them in some way, then there may be a tendency towards slow movement. Aphids, such as greenflies and blackflies, feeding on the juices of plant stems, usually have all the food they need. They are so easily caught and eaten by their enemies, *e.g.* ladybirds, that only their vast numbers ensure the continuation of their kind. A number of animals living in fresh and salt water have their food conveyed to them by currents of water and tend to become fixed in one place. The mussel and the limpet, for example, have their food conveyed to them. Their shells afford them protection and they have little need to travel.

Just as animals may move to escape their enemies, so they may also move to find suitable conditions for reproduction. Birds can use their wings to escape their enemies; they can also fly to inaccessible places to lay their eggs and to environments where the better rearing of the young can be undertaken. Migration is associated with breeding and rearing as well as with sources of food.

It is only amongst insects and birds that flying is widespread. Some pre-historic reptiles – the pterodactyls – were able to fly, but they became extinct. Of the mammals, only bats can fly. Other animals, such as the squirrel and the flying fish, are able to glide for short distances, but they cannot actually fly, nor can they change direction when in the air.

The evolution of animals which were able to live and feed on the land was accompanied by the evolution of methods of obtaining oxygen from the air instead of from water. These included the development of lungs in amphibians, reptiles, birds and mammals, and the development of air tubes opening on to the body surfaces of insects. However, as time went by, individual members of the various classes of land-living animals returned to water to feed. In such animals the physical structure has often altered in some way to assist them in forcing themselves through the water, but in most cases they are still faced with the necessity of obtaining their oxygen from the air. Whales are good examples of this.

The four classes of animals introduced in Pupils' Book 1 were mammals, birds, insects and fish. The illustrations for Lesson 7 in Pupils' Book 2 show:

1   a representative of each of these four classes in water
2   a representative of each of these four classes on land
3   a representative of each of these four classes in the air

### In Water

*Mallard Duck.* This is an example of one of the many birds which have webbed feet to enable them to force themselves through the water. It is at home on land, in air and in water. In water it feeds on both animal and plant foods sifted from the mud. On land it feeds on seeds and such small animals as insects and worms. It is resident throughout the British Isles and is perhaps the best known of our wild ducks.

*Freshwater Beetle (Dytiscus marginalis).* This insect is another animal which can move from place to place on land, in water and in the air. It is typical of an insect which has returned to spend most of its life in water, although both as a larva and as an adult it has to come to the surface to obtain its oxygen from the air. It feeds on other animals in the pond – even its own kind – and, being a powerful swimmer, can catch small fish. The hind legs are adapted for use in swimming, and the front legs are used for holding prey. When it does travel from one pond to another, it goes at night, crawling out on to the land and then using its wings.

*Swordfish.* The swordfish is amongst the swiftest of fish with a possible speed of about 95 kilometres per hour. Its nose is drawn out into a flat, sharp-edged sword which may grow to about 6 metres. Some say that this is used with a flail-like motion to kill or wound other fish, especially those in shoals, and that the point is used to protect it from larger animals in the ocean. Other authorities, however, doubt that it is ever used for offensive purposes.

*Whale.* Whales are mammals which have returned to water. There are two varieties – the whalebone (or baleen) whales and the toothed whales. The latter include the killers, the dolphins, sperm whales and porpoises. The fore limbs are reduced to flippers, and hind limbs are no longer visible externally. As they still breathe by means of lungs, they have to come periodically to the surface, and their 'blowing' is due to the exhaled breath forcing up a cloud of spray. Food varies from minute aquatic life to large animals, according to the species. So far as is known, blue whales are the largest animals ever to have lived.

### On Land

*Cockroach.* Cockroaches are believed to be the fastest insects on land, using their legs to carry them swiftly to the dark places which they prefer. Wings may be used for flying, but they are sometimes absent, especially on females. Their objectionable smell and the fact that they feed on a variety of animal and plant foods makes them a nuisance to human beings. Although often called 'black beetles', they are neither black, nor are they beetles.

*Kiwi.* Like the ostrich and the penguin, this bird no longer uses its wings for flying and relies on its ability to run swiftly. It is resident in

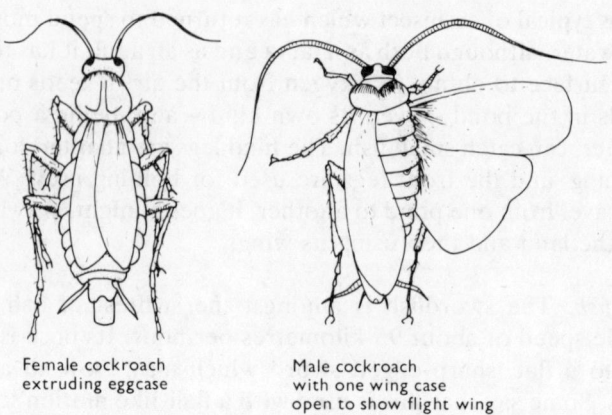

Female cockroach
extruding eggcase

Male cockroach
with one wing case
open to show flight wing

New Zealand, where it is nocturnal, feeding mainly on worms.

*Mole*. With its strong fore-legs adapted for burrowing, this animal lives mainly below ground. Food is strictly animal – mainly worms, insects and their larvae, although flesh from dead birds, mammals and reptiles is also eaten. The mole wriggles awkwardly on the surface, but, in water, is a good swimmer. Most of its burrowing is in the constant search for food. A mole hill is the main nest; smaller mole hills are thrown up during burrowing.

*Eel*. Most of the common eel's life is spent in fresh water, but it is born in the Sargasso Sea, and it returns there to lay its eggs. Where conditions are suitably damp, it can travel over land from one section of water to another. Food consists mainly of a variety of other small animals.

## In the Air

*Flying Fish*. In spite of its name, this fish cannot fly. When escaping from enemies, it can soar into the air and glide – assisted by its enlarged pectoral fins – up to 140 metres. Being unable to change direction in the air, these fish sometimes land on vessels. Food consists of smaller fish.

*Swift*. This summer visitor to Britain is one of the swiftest birds in the air, feeding on insects in flight. Its short legs are virtually useless for walking.

*Bat.* Apart from birds and insects, this mammal is the only animal which can fly. Bats have no true wings, but merely a membrane of skin stretched between fore and hind limbs. Nocturnal flyers, they feed mainly on insects. The echoes from the supersonic sounds which they emit guide their flight, acting like a radar system and enabling them to avoid obstacles.

*Dragonfly.* This insect passes through a three-stage life, the first two stages being spent in water, where the larva feeds on small aquatic animals. The adult – confined to land and air – is believed to be the fastest insect in the air, where it feeds, like the swift, on insects caught in flight.

## CODE

1 Observe the different ways in which animals move from place to place, *e.g.* crawl, walk, run, swim, fly, glide.
2 Observe how some animals have limbs to assist in movement and others have not.
3 Observe that some birds walk and others hop.
4 Observe that some birds have webbed feet for swimming.
5 Observe that some animals have adapted their capability of movement to more than one medium, *e.g.*
  *a* humans can move on land and in water
  *b* most kinds of birds, and most kinds of adult insects can move on land and in the air
  *c* some insects, *e.g.* the adult water beetle, and some birds, *e.g.* ducks, swans and gulls, can move on land, in water, and in the air.
6 Observe that some birds do not fly much, *e.g.* domestic fowls.
7 Observe the autumn migration of birds.
8 Observe animals that can move in more than one medium (water and land, *or* land and air, *or* water, land and air).

### Written Work

1 Living animals move to find their <u>food</u>.
2 <u>Whales</u> are mammals which live in the sea.
3 <u>Bats</u> are mammals which fly in the air.
4 A living animal moves by its own <u>effort</u>.

# 8 OTHER THINGS ARE MOVED FROM PLACE TO PLACE

## BY MOVING ANIMALS, MOVING GASES AND MOVING LIQUIDS

### Demonstration Material

1 Any lightweight object which floats on water, *e.g.* table tennis ball, cork, ball of paper, etc.
2 Bowl of water.
3 Any fruits with adaptations for being carried by animals, water or wind.

### Sample Link Questions

1 What are the three forms in which we find never-alive things? (*Solid, liquid, gas*)
2 Which of these three has both size and shape? (*The solid*)
3 Which has size but no shape of its own? (*The liquid*)
4 Which has neither size nor shape of it own? (*The gas*)
5 What is the difference between living animals and living plants? (*Living animals can move from place to place*)
6 What are the two main reasons for which living animals move from place to place? (*To find food and to escape their enemies*)
7 Which two classes of animals have wings? (*Birds and insects*)
8 Which are the only mammals which can fly in the air? (*Bats*)
9 What is air a mixture of? (*Gases*)
10 What name do we give to air when it is moving? (*Wind*)

### Relevant Information

The main points of this lesson are:
1 living, dead or never-alive things can be *forced* to move from place to place by something else which is moving
2 the three main agents responsible for moving things from one place to another are:
*a* moving animals
*b* moving air – wind
*c* moving water

3 of these three, only living animals can force themselves to move. Both moving air and moving water are in turn forced to move.

*Note*

This lesson serves as an introduction to forces, and the word *force* is emphasised throughout.

## Moving Animals

Just as living animals can force themselves to move from place to place, so they can force other things to move from place to place. In general, animals carry

*a* food – either for themselves or for their young
*b* their young – to protect them from enemies
*c* building material – *e.g.* birds carry material for nest-building

Most animals use their mouths or their limbs for carrying things. There may even be adaptations to assist in this. The golden hamster, for example, has enlarged cheek pouches. Some birds use their beaks, and others, like the eagle, use their feet. The kangaroo is an example of an animal with a specially developed pouch for carrying its offspring.

## Moving Gases

The commonest gas (or rather mixture of gases) is, of course, air. When a surface becomes hot, the air above it also becomes hot, expands, becomes lighter, and is forced to rise, as gravity forces colder, heavier currents to move in. The resulting wind may force many other things to move from place to place.

## Moving Liquids

Any moving liquid is capable of forcing something else to move. Water is, of course, by far the commonest liquid in the world and is particularly responsible for carrying material from the land towards the sea. The waters of streams or rivers are themselves forced to move towards the sea by gravity.

Anything in motion may carry something else. Glaciers are well-known examples of moving solids which carry material – particularly rocks – and during the ice age many rocks were carried and deposited by glacial action.

Non-microscopic plants have no need to move from place to place, but it is necessary for the propagation of the species that their spores or seeds should be scattered in some way. Some plants have evolved special adaptations on their fruits or seeds to enable them to be carried by moving animals, moving air, or moving water.

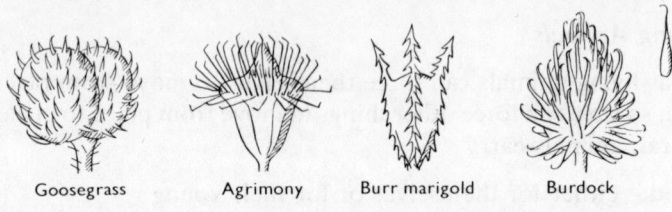

Goosegrass          Agrimony          Burr marigold          Burdock

FRUITS ADAPTED FOR CARRIAGE BY ANIMALS

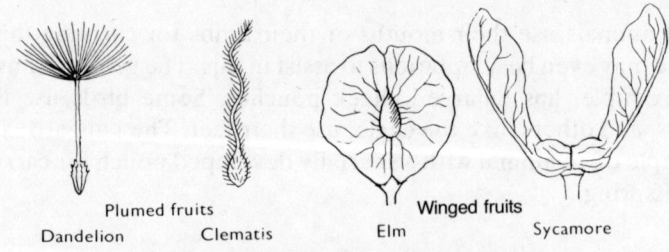

Plumed fruits                    Winged fruits
Dandelion          Clematis          Elm          Sycamore

FRUITS ADAPTED FOR CARRIAGE BY AIR

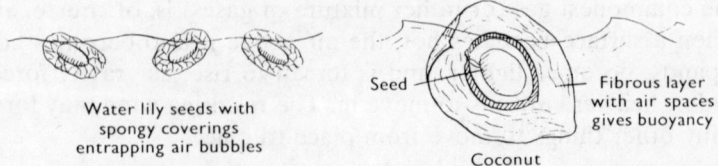

Water lily seeds with spongy coverings entrapping air bubbles

Seed

Fibrous layer with air spaces gives buoyancy

Coconut

SEEDS AND FRUIT ADAPTED FOR CARRIAGE BY WATER

## CODE

1  *a* Find a way of forcing a table tennis ball to move from one place to another place without touching it.

   *b* Demonstrate that the object can be forced to move from one place to another if somebody blows it.

    *c* Observe that it is moving air which forces the ball to move.

    *d* Observe that the air itself is forced to move by whoever does the blowing.

2   *a* Float a table tennis or pith ball on water, and find a way of forcing it to move across the surface without being touched.

    *b* Demonstrate that the ball can be forced to move across the surface if the water is stroked in one direction with a hand or paddle.

    *c* Observe that it is moving water which forces the object to move.

    *d* Observe that the water itself is forced to move by whoever does the 'paddling'.

*Note*

For both 1 and 2 any other suitable floating object, such as a cork or ball of paper, would suffice.

3   Observe from things which are carried (*e.g.* clouds and autumn leaves) that there is moving air at different heights above the ground.

4   Collect and observe any fruits or seeds with adaptations for being carried by *a* animals, *b* wind, *c* water.


**Written Work**

1   A living <u>animal</u> chooses when to move and when to stop.

2   Other things <u>move</u> when they are forced to.

3   Many things are moved by animals, air and <u>water.</u>

4   Animals may carry things for <u>food</u> and to build with.

5   Air and water move only when they are <u>forced</u> to.

# 9 FORCING THINGS TO MOVE

## HOW WE MOVE THINGS FROM PLACE TO PLACE ON LAND

### Demonstration Material

1   Two children, or one child and a heavy weight.
2   Two matchbox covers, or two pieces of sandpaper.
3   A round tin and a board or a book.
4   Demonstration model to show that sliding is easier than carrying, and that rolling is easier than sliding (see CODE 5).

### Sample Link Questions

1   What is the difference between living animals and plants? (*Living animals can move from place to place*)
2   What are the three main things which can carry others from place to place? (*Moving animals, moving liquid, and moving gas – animals, air and water*)
3   Of these, which are the only ones which can force themselves to move? (*Living animals*)
4   Water flows and wind blows, but do they force themselves to move? (*No, they are forced to move*)
5   Which are the only things that can choose what they carry? (*Living animals*)
6   What are the two main uses which animals have for the things they carry? (*Food and building*)
7   Why are liquids and gases easily forced to move? (*They have no shape to stop them flowing*)

### Relevant Information

The main purpose of this lesson is to show that the three main ways in which we can force things to move from place to place on land are by:

1   carrying
2   sliding (which may be caused by pushing or pulling)
3   rolling.

In their movement in water, on land or in the air, all living animals are impeded by two major forces, namely gravity and friction. These same two forces need to be overcome when other things are moved from place to place. One of the ways in which human beings have elevated themselves above the rest of the animal kingdom is by developing more efficient methods of overcoming these two forces.

On the land itself, living animals resort to many different ways of forcing other things to move from place to place. These different ways may be grouped under the three general headings of carrying, sliding and rolling, and before the inception of vehicles, they were probably resorted to in that order. They serve as an introduction to the methods by which human beings overcome gravity and friction. Gravity is considered in this book and friction in books 3 and 4.

*Carrying*

Carrying is the most obvious method adopted by animals in general, mouth and limbs being used in various ways for this purpose. Human beings carry things in their hands, under their arms, on their backs, over their shoulders, and on their heads. Shoulders and heads offer a direct support, and are therefore used for supporting heavy burdens once they have been lifted. Arab women carrying pitchers of water are one example of people who use their heads; African porters carrying packages in this way, and market porters carrying tiers of baskets are others.

Before a weight can be carried, it has usually to be lifted. That is to say the force of gravity has to be directly overcome. It follows, that if the force of gravity can easily be overcome – that is if the object to be moved can be lifted and supported easily – then carrying is a suitable means of transporting it. Pens, pencils, books, and the wide variety of things which small boys carry about in their pockets are obvious examples of things which human beings find suitable for carrying. However, where the weight of an object is such that it is difficult to support it against the force of gravity for long, or where its weight is such that it cannot even be lifted, then other methods of moving it need to be found.

*Sliding*

There are two obvious ways of forcing things to slide:

*a* pushing, *i.e.* applying the force behind
*b* pulling or dragging, *i.e.* applying the force in front

97

In sliding, a force of resistance (friction) is set up between the surfaces in contact. The rougher the surface, the greater the friction, and the harder it is to force the thing to slide. The smoother the surfaces in contact, the less the friction, and the easier it is to force the thing to slide. Dragging a sledge over a smooth frozen surface is easier than dragging it over rough ground. Over ice and snow a sledge can be better than a wheeled vehicle. This is when the friction is so considerably reduced that turning wheels only slide against the smooth surface instead of obtaining a grip.

The friction between two surfaces which are in contact is proportional to the load – proportional to the force pressing them together – and is independent of the area of surface which is in contact. Provided the load remains the same, the friction will remain the same, regardless of the area of surface in contact. (This assumes, of course, that the smoothness of the surface in contact remains the same.) Consequently the same force will be required to slide a weight, no matter which way up it stands. This is illustrated below.

THE SAME FORCE IS REQUIRED TO SLIDE A WEIGHT NO MATTER WHICH WAY UP IT STANDS

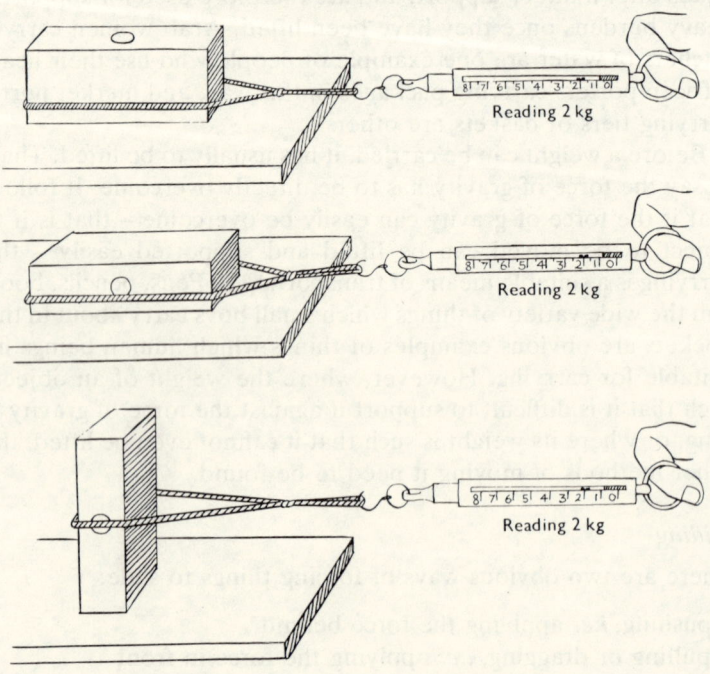

98

## Pulling and Pushing

When sliding an object, it is usually easier to pull it than to push it. This is because there is a tendency in pulling to lift the object up from the ground, whereas in pushing, there is a tendency to push it downwards towards the ground.

PULLING MAY BE EASIER THAN PUSHING

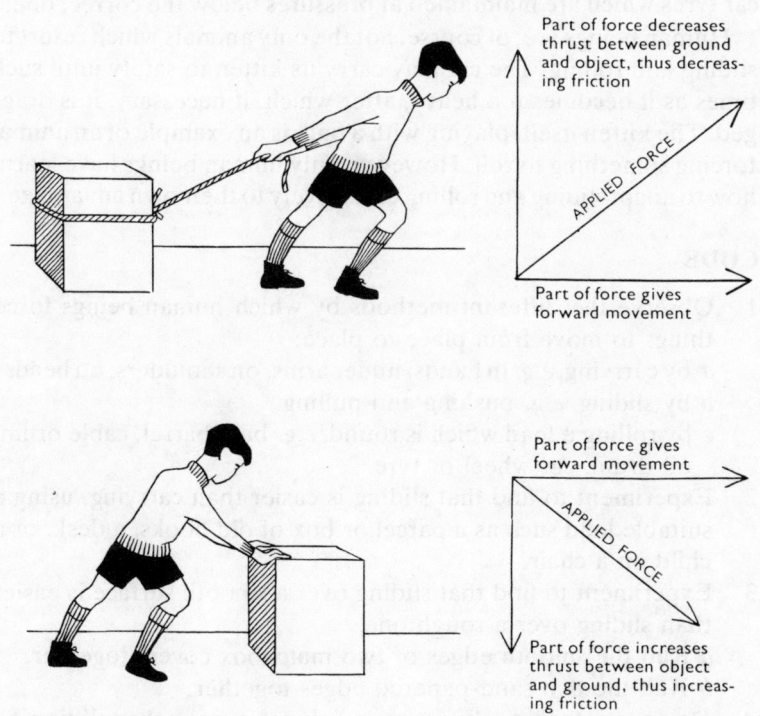

Part of force decreases thrust between ground and object, thus decreasing friction

APPLIED FORCE

Part of force gives forward movement

Part of force gives forward movement

APPLIED FORCE

Part of force increases thrust between object and ground, thus increasing friction

## Rolling

When the object is round, the easiest method of moving it from one place to another is by forcing it to roll. In rolling, little effort is required to overcome friction. The point of contact between a rolling object and its carrying surface is momentarily at rest. This means that in theory there is no friction of movement between the two surfaces.

However, the static friction (*i.e.* the friction between two surfaces which are not in motion across one another) must be sufficient to ensure that this momentary position of rest is maintained. If the static friction is insufficient, then sliding (by slipping, or skidding) will take place.

In everyday practice either a flattening of the rolling object or an indentation of the surface at the point of contact takes place. Owing to this, a certain amount of sliding is produced. It is this sliding friction which causes wear. This is especially noticeable on motor-car tyres which are maintained at pressures below the correct one.

Human beings are, of course, not the only animals which resort to sliding and rolling. The cat may carry its kitten to safety until such times as it becomes too heavy, after which, if necessary, it is dragged. The kitten itself, playing with a ball, is an example of an animal forcing something to roll. However, only human beings have learnt how to adapt sliding and rolling extensively to their own advantage.

## CODE

1 Observe the different methods by which human beings force things to move from place to place:
   *a* by carrying, *e.g.* in hands, under arms, on shoulders, on heads
   *b* by sliding, *e.g.* pushing and pulling
   *c* by rolling a load which is round, *e.g.* ball, barrel, cable drum, oil drum, car wheel or tyre.
2 Experiment to find that sliding is easier than carrying, using a suitable load such as a parcel or box of old books, a desk, or a child on a chair.
3 Experiment to find that sliding over a smooth surface is easier than sliding over a rough one.
   *a* Rub the smooth edges of two matchbox covers together.
   *b* Rub the two sand-papered edges together.
4 Demonstrate that rolling a round object is easier than sliding it.
   *a* Stand a cylindrical object (say, a round tin) on a book or a length of wood and tilt until the object slides down the sloping surface.
   *b* Now place the cylinder on its side and again tilt the surface. Only a slight tilt will make it roll.
5 Demonstrate with one object that sliding is easier than carrying and that rolling is easier than sliding.

## Preparation

1 Fill a cylindrical tin with sand, and fix the lid securely.

2 Tie a length of string (or thin wire) round the tin, and fix in position with tape. Leave a gap to allow a paper clip to be attached.

3 Fasten a paper clip to each end of a separate length of string (or wire). This should be long enough to wind round the tin three or four times.

4 Fasten another paper clip to a length of elastic, or a suitable elastic band.

## Demonstration

### Carrying

1 Lay the tin on its side.

2 By means of the paper clip attach the elastic or elastic band to the string wound round the tin.

3 Lift up the tin.

4 Measure with a ruler the length to which the elastic is stretched.

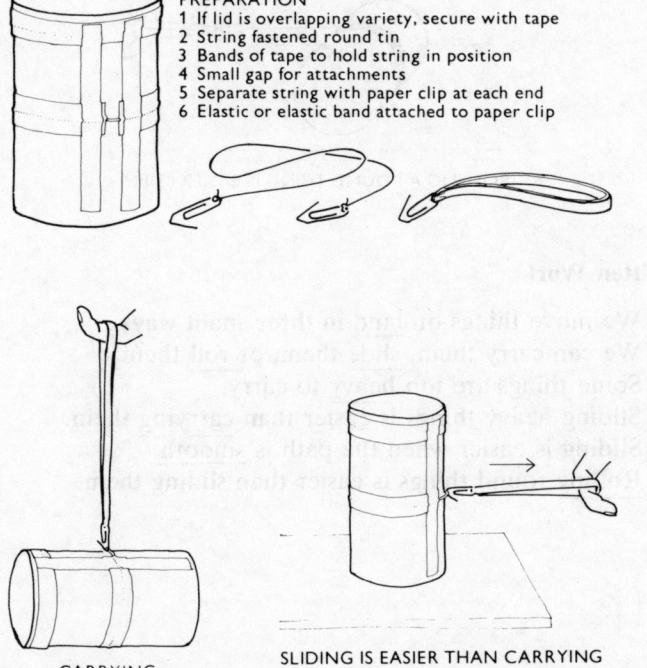

PREPARATION
1 If lid is overlapping variety, secure with tape
2 String fastened round tin
3 Bands of tape to hold string in position
4 Small gap for attachments
5 Separate string with paper clip at each end
6 Elastic or elastic band attached to paper clip

CARRYING

SLIDING IS EASIER THAN CARRYING

*Sliding*

1 Stand the tin on end.
2 By means of the elastic, start to pull the tin along a desk or table top.
3 Measure with a ruler the length to which the elastic is stretched, *while the tin is sliding*.

*Rolling*

1 Remove the elastic and its paper clip and in their place attach the separate length of string.
2 Wind the loose end of the string round the tin, and lay the tin on its side.
3 Attach the elastic to the paper clip at the free end of the string.
4 Pull the elastic until the tin begins to roll.
5 Observe (or measure with a ruler if possible) the very slight extent to which the elastic is stretched *while the tin is rolling*.

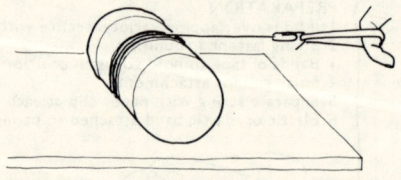

ROLLING A ROUND THING IS EASIER STILL

**Written Work**

1 We move things on <u>land</u> in three main ways.
2 We can carry them, <u>slide</u> them, or <u>roll</u> them.
3 Some things are too <u>heavy</u> to carry.
4 Sliding heavy things is easier than <u>carrying</u> them.
5 Sliding is easier when the path is <u>smooth.</u>
6 <u>Rolling</u> round things is easier than sliding them.

# MOVING THINGS FROM PLACE TO PLACE ON LAND

## ROLLERS AND WHEELS HELP

### Demonstration Material

1 A brick with string or wire fastened round it. (See CODE 1).
2 Elastic and paper clip – as used in Lesson 9.
3 Rollers, *e.g.* round pencils, lengths of dowel, or lengths of broom handle.
4 Any simple non-mechanical toys employing wheels and axles, *e.g.* toy motor-cars.

### Sample Link Questions

1 How is it that never-alive things sometimes move from place to place, even though they cannot move themselves? (*They are forced to*)
2 What are two never-alive things which, when they are forced to move, often force other things to move from place to place at the same time? (*Moving gas – wind – and moving liquid – water*)
3 What are three main ways in which *we* can force other things to move from place to place on land? (*Carrying, sliding, rolling*)
4 Why is it that we cannot carry some things? (*They are too heavy*)
5 What is usually the easiest way of moving heavy things – by carrying or by sliding? (*Sliding*)
6 How do we force things to slide? (*Push, pull or drag*)
7 What kind of path makes it easier to slide things – rough or smooth? (*Smooth*)
8 What shape have things to be before we can roll them? (*Round*)
9 Which is the easier way of moving heavy round things – sliding or rolling? (*Rolling*)
10 Why can we not roll everything which is too heavy to carry? (*Everything is not round*)

### Relevant Information

The main purpose of this lesson is to show:

1 how the method of rolling can be used to force things to move when they themselves are not round

103

2   the development of simple wheels and axles from rollers
3   the use of different numbers of wheels on land vehicles.

The different methods of forcing wheels to turn are dealt with in separate lessons in Books 3 and 4.

Forcing a round object to move from one place to another by causing it to roll is not a particularly outstanding achievement. Using a round object as an aid when forcing some other object to move, is. The first prehistoric man to do this placed the human mammal ahead of all other animals. It is not known who first used rollers, or when, but it is logical to assume that the first type of roller to be used was the stem of a tree, as tree trunks would be the most likely rollers to be found naturally. It is also believed that rollers were first used in the stone age.

Where several tree trunks were used as rollers for moving heavy weights, there would be obvious disadvantages.

a It would be unlikely that all the rollers would be the same diameter.
b Progress would be slow, because of the constant necessity for moving the rear rollers to the front.
c The rollers themselves would probably be heavy and cumbersome.

These difficulties may have been overcome to some extent, by thinning the trunk, except at the ends, so as to leave a central stem with a disc at each end. The advantages would be:

a to make the roller itself lighter
b to make it easier to keep the load in place
c that only the discs at the ends would be in contact with the ground, and the greater part of the roller would be kept clear of obstacles.

This was possibly a first step towards the development of wheels. Although the roller would be lighter and easier to handle, movement would still be slow as the rear roller still had to be moved to the front. Alternative solutions were:

1   to fasten the load in some way to the axle of the roller so that the axle was still free to turn. This of course would lead to much friction between the axle and the load.
2   to make only the discs turn, while the axle and load remained fixed. This meant that the discs had to be made separately, each

with a hole in the centre which would fit over the end of the axle. And so the wheel was invented.

The invention of wheels, which at first were probably slices of tree trunks, obviously allowed heavy loads to be carried more quickly. Where these slices of tree trunk were heavy, they in turn may have been thinned in the middle to make them lighter.

The need for strength plus lightness plus similarity of size probably led to the next development – spokes.

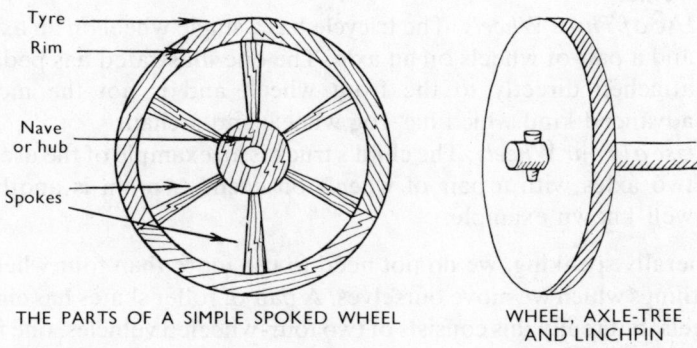

THE PARTS OF A SIMPLE SPOKED WHEEL

WHEEL, AXLE-TREE
AND LINCHPIN

The full name of the bar on which a wheel turns is the *axle-tree* (as on a cart). It is very often referred to as the axle, but this particularly applies when the bar itself revolves in bearings and has the wheels fixed to it.

Wheels were prevented from falling off the axle-tree by a *linchpin*, which fitted through a hole in the end of the axle-tree. Nowadays, of course, better methods fulfil this function.

It is believed that wheels were invented some 4,000 to 5,000 years ago by the primitive Indo-European people. The spoked wheel possibly originated in Egypt, where there would be plenty of flat ground suitable for the use of wheels. Illustrations of spoked wheels have been found on tombs in Egypt. The number of spokes on these illustrated wheels is either four or six.

The illustrations for Lesson 10 in the Pupils' Book show:

A The development of an early axle and wheels from primitive rollers.
B Examples of a modern roller, and wheeled vehicles designed for being forced to move by human muscle power alone.

1   *Use of a Single Roller or Wheel*. In a garden roller the full weight of the whole roller itself is required. The wheelbarrow on the other hand is a one-wheeled vehicle designed to enable weights to be moved, so that the wheel needs to be as light as possible.
2   *Use of Two Wheels*. Two wheels are shown in use in different ways.
    *a* The scooter has two single wheels set one behind the other, each on its own axle.
    *b* The truck has a pair of wheels set at opposite ends of the same axle.
3   *Use of Three Wheels*. The tricycle has a single wheel on an axle, and a pair of wheels on an axle. The one illustrated has pedals attached directly to the front wheel, and is not the more advanced kind which has cog wheels and a chain.
4   *Use of Four Wheels*. The child's truck is an example of the use of two axles with a pair of wheels on each. A pram is another well-known example.

Generally speaking, we do not need to use more than four wheels for things which we move ourselves. A pair of roller skates has eight wheels, but really this consists of two four-wheeled vehicles, one for each foot.

Two points worthy of note here are that

1   although the invention of wheels may have resulted from the need to move heavy weights from one place to another, many uses have been found for wheels in addition to their uses for transport;
2   although turning wheels are invaluable to us in modern civilisation, they still have to be forced to turn, and finding sources of power capable of turning the wheels for us always has been – and still is – one of the problems of mankind.

It is obvious that if the wheel had never been invented, modern civilisation as we know it could never have developed.

**CODE**

1   Demonstrate that an object which is not round can be forced to move more easily on rollers than by sliding it.

**Preparation**

*a* Tie a length of string (or thin wire) round a brick and fix in position with tape. Leave a gap at one end to allow a paper clip to be attached.

*b* Acquire a number of rollers, *e.g.* round pencils, or lengths cut from a broom handle or a dowel rod.

*c* Fasten a paper clip to a length of elastic, or a suitable elastic band (or use the elastic and paper clip from the previous lesson).

## Demonstration

*Forcing a load to move by sliding*

*a* By means of the paper clip, attach the elastic or elastic band to the string wound round the brick.

*b* Begin to drag the brick.

*c* Measure with a ruler the length to which the elastic is stretched *while the brick is sliding*.

*Forcing the load to move on rollers*

*a* Mount the brick on a number of rollers.

*b* By means of the elastic, pull the brick, so that it begins to move on the rollers.

*c* Observe (or measure with a ruler if possible) the very slight extent to which the elastic is stretched *while the brick is moving*.

*d* Observe the disadvantage of having to move rollers from back to front repeatedly.

2 Observe the number of wheels in use on wheeled vehicles.

3 Observe how wheels are fixed on toy motor-cars and other children's toys.

4 Observe how wheels are fixed on any models made up from children's construction sets.

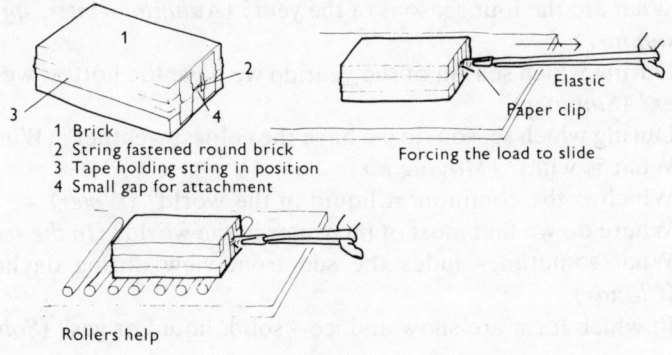

1 Brick
2 String fastened round brick
3 Tape holding string in position
4 Small gap for attachment

Elastic
Paper clip
Forcing the load to slide

Rollers help

**Written Work**

1 <u>Heavy</u> loads were moved ages ago on rollers.
2 A light roller is <u>easier</u> to turn than a heavy one.
3 When only the ends turn, we call them <u>wheels</u>.
4 The bar on which wheels turn is called an <u>axle</u>.
5 Rollers and wheels have to be <u>forced</u> to turn.

# THE THREE WEATHER MAKERS

## SUN, AIR AND WATER

**Demonstration Material**

1 Saucer or tin lid.
2 Mirror, or clean jam jar, or milk bottle.
3 Tall clear glass jar and gummed paper strip.
4 Two cloths of the same material and size.
5 Blackboard.
6 Water.
7 Stiff paper or card.

**Sample Link Questions**

1 What gives us warmth outside our homes? (*The sun*)
2 How long does it take the earth to go once round the sun? (*One year*)
3 What are the four seasons of the year? (*Autumn, winter, spring, summer*)
4 During which season of the year do we have the hottest weather? (*Summer*)
5 During which season do we have the coldest weather? (*Winter*)
6 What is wind? (*Moving air*)
7 Which is the commonest liquid in the world? (*Water*)
8 Where do we find most of the water in the world? (*In the seas*)
9 What sometimes hides the sun from view during daylight? (*Clouds*)
10 In which form are snow and ice – solid, liquid or gas? (*Solid*)

## Relevant Information

Weather is a happening. The purpose of this lesson is to show that the three main factors contributing to the weather are:

1 sun – which gives us heat
2 air – which moves at different speeds
3 water – in one or more of its forms.

The warmth of the sun, the movement of air and the presence of water in one form or another are the three factors which, between them, are responsible for the state of affairs which we call 'weather'.

The heat from the sun not only provides us with warmth; it is also responsible for

1 forcing air to move (the formation of air currents)
2 the evaporation of liquid water into water vapour.

### The Sun and the Air

When a surface becomes heated, the air above it becomes heated, expands, and becomes lighter. As liquids and gases are things with no fixed shape, they tend to flow downwards due to the earth's gravity, and naturally the heaviest will settle at the bottom. Thus, when air is heated and becomes lighter, it is forced to move upwards by the surrounding colder and heavier air flowing in to replace it. It is because of this that we have the concise but rather misleading statement that hot air rises. It does of course rise, but only because it is forced to.

Winds are usually measured in kilometres per hour, or in knots, and are given a notation from the Beaufort Scale.

According to the direction from which they have come, winds may be warm or cold, wet or dry. Here are some special names by which some winds are known.

*Hurricane* – the strongest of winds. The term was originally used in the West Indies, and was derived from Hunraken, the Carib Indians' god of storms. Hurricanes die out over land.

*Typhoon* is the name by which a hurricane is known in the China Seas.

*Cyclone* is the name by which a hurricane is known in the Indian Ocean.

109

*Willy-Willy* is the name by which a hurricane is known off the north-west coast of Australia.

*Tornado* – a storm wind of the southern states of the U.S.A. Warning is given by ominous clouds with tinted clouds moving round them. It has a small front, but the whirling wind of over 160 kilometres per hour causes a column of cloud to descend like an enormous trunk. The centre of the 'trunk' is a partial vacuum, and this plus the whirling wind causes great destruction. When a similar phenomena occurs over the sea, it is called a waterspout, but the sea water is drawn up for only a short distance. Rain which falls in a waterspout is not sea water which has been forced up, but fresh water. Tornadoes have occurred in Britain.

*Sirocco* – a hot dry wind which carries sand with it. Such sandstorms occur in desert regions, particularly the Sahara.

## The Sun, the Air, and Water

Water is a liquid, and it is visible. Water vapour is a gas, and it is invisible.

Although water vapour is one of the gases in the air, it is present only in very small amounts. It has been estimated that the water vapour content varies from 0.01% to 4%. The evaporation of liquid water into water vapour takes place at the surface of the liquid and is influenced by the temperature.

Evaporation continues so long as the air above the surface does not become saturated. Evaporation is assisted if the air in contact with the surface is moving.

As warm air is forced upwards into regions of lower temperature, it may reach a level where the temperature is low enough for the water vapour to condense back into tiny drops of liquid. These tiny visible drops are the beginnings of cloud formation.

*Clouds* occur at various heights, and the highest, the cirrus clouds, occur at an average height of 9 kilometres and may consist of ice crystals. When the tiny droplets of water which go to form a visible cloud, are cooled even more, they form larger drops which, if sufficiently heavy, fall to earth.

*Fog* is a cloud of droplets of water at ground level. It is in effect a cloud on the ground and is brought about by the cooling of water vapour near to the ground, and the resulting condensation into liquid. Thin fog is called *mist*. The thick fog of cities is formed by the condensation into droplets taking place on the tiny solid particles of soot and other by-products of smoking chimneys. 'Smog' is a coined word used to describe the kind of fog which is made up partly of the impurities which are poured into the air.

It has been estimated that a thick fog contains 1 500 to 2 000 droplets of condensation per cubic centimetre, and even then the droplets are so tiny that the cubic centimetre is not overcrowded. This is an indication that, when water vapour condenses into liquid, the droplets of liquid are very small.

The foggiest area in the world is accepted as being the Grand Banks of Newfoundland which experience a considerable amount of sea fog.

*Snow* is formed by water vapour in the air encountering temperatures below freezing point, so that, instead of condensing into liquid droplets, it freezes into solid crystals which combine to form snowflakes. Snowflakes are therefore made up of tiny ice crystals, and these vary greatly in the pattern they form, so that no two snowflakes can be found alike. The ultimate shape however, is almost always that of a hexagon.

*Hail*. The formation of hailstones takes place in stages. First, a little water condenses on a tiny solid particle in the air. This, upon meeting other particles, grows in size. The resulting droplet falls towards the ground. Second, a strong ascending current of air carries the raindrop aloft. If the level to which the droplet is carried by the updraught has a sufficiently low temperature, the droplet will freeze and fall as a hailstone. It is during summer thunderstorms when there are violent updraughts that this usually happens. In violent thunderstorms, the falling hailstones may be carried up again to form a further layer of ice. This may happen repeatedly, and the hailstone increases in size with each successive layer of ice formed.

Hailstones in Britain and Europe are not normally larger than peas, although they have been known to attain the size of golf balls. It is in the sub-tropics that they are more likely to attain the size of tennis balls.

*Sleet* can be formed in two ways:

*a* snow falling from a high altitude may partly melt on the way down
*b* rain falling from a high altitude may partly freeze on the way down.

*Notes*

1  The weather varies from day to day and from locality to locality. The average weather of a particular locality, taken over a period of years, is known as its climate.

2  The idea that the weather on St. Swithin's Day (15th July) will determine the weather for the next forty days is only a legend.

## CODE

1  Observe how, as the hours of daylight grow shorter, the weather becomes colder.

2  Observe that winds blow from different directions and at different speeds.

3  Observe that washing dries more quickly
   *a* on hot days than it does on cold days;
   *b* on windy days than it does when the air is still.

4  Observe the drops of water that collect on clothing when walking through fog.

5  Demonstrate how tiny drops of water form from invisible water vapour in the air by breathing
   *a* on to a mirror;
   *b* through a straw into a clean jam jar or milk bottle.

6  Demonstrate that water disappears into the air as water vapour, by leaving a saucer or tin lid full of water in some part of the room.

7  Experiment to show the rate at which water in the classroom disappears as water vapour.
   *a* Fasten a strip of gummed paper from top to bottom of a tall jar.

   *b* Almost fill the jar with water, and make a mark on the paper to indicate the surface level.

112

    *c* Mark the new level each week, to show how much water has evaporated.

8 Experiment with a control to show that heat helps water to change to water vapour.

    *a* Soak in water two pieces of cloth of similar material and equal size.

    *b* Place one on a warm radiator, and the other well away from all heat.

9 Experiment with a control to show that moving air helps water to change into water vapour.

    *a* Mark off two equal areas at opposite ends of a blackboard (away from heat).

    *b* Wet each area with a damp cloth.

    *c* Fan one with a piece of stiff cardboard.

10 Observe and record the weather daily, under the headings of SUN, AIR, WATER.

    *a* Under the column headed SUN use terms such as 'Sunny', 'Not sunny'.

    *b* Under the column headed AIR use terms such as 'Calm', 'Breezy', 'Windy'.

    *c* Under the column headed WATER use terms such as 'Rain', 'No rain', 'Snow', 'Hail', 'Sleet', 'Frosty'.

| DAY | DATE | SUN | AIR | WATER |
|-----|------|-----|-----|-------|
| MONDAY | | Not sunny | Breezy | Rain |
| TUESDAY | | Sunny | Calm | No rain |
| WEDNESDAY | | Not sunny | Windy | No rain |
| THURSDAY | | | | |
| FRIDAY | | | | |

**Written Work**

1 The three weather makers are the sun, air and water.

2 Sun and wind help water to become water vapour.

3 Water vapour is a gas in the air.

4 Tiny drops of water form when water vapour cools.

5 Moving air forces the clouds to move.

6 Mist and fog are clouds near to the ground.

113

# FORCING THINGS TO MOVE

## THE EARTH PULLS

### Demonstration Material

1  Suitable load, *e.g.* a brick with string round it, as used in Lesson 10.
2  Elastic and paper clip as used in Lesson 9 and 10.
3  Heavy toy vehicle on wheels, or rollers as used in Lesson 10.
4  A board and some books to make a slope.
5  'Plumb bob', *e.g.* a weight suspended on a thread from a ruler, or a conker on a string.

### Sample Link Questions

1  Name a liquid which carries many things from place to place when it is moving. (*Water*)
2  Name a mixture of gases which carries many things from place to place when it is moving. (*Air*)
3  What are three ways in which we force other things to move from place to place on land? (*Carrying, sliding, rolling*)
4  When forcing heavy things to move from place to place, which is the hardest way – carrying, sliding, or rolling? (*Carrying*)
5  What do we use to help us to move things which are not round and which are too heavy for carrying or sliding? (*Rollers and wheels*)
6  What are the three weather makers? (*Sun, air, water*)
7  When water dries up, what does it become? (*Water vapour*)
8  When water vapour cools what is formed? (*Water*)
9  What is a cloud made of? (*Tiny drops of water*)
10  What forces the clouds to move? (*Moving air – wind*)

### Relevant Information

The main points of this lesson are:

1  earth has a pull which can force things to move. We call this pull *gravity*
2  because of gravity:

    *a* things are forced to fall (towards the earth's centre)
    *b* downhill is easier than uphill
3   when we lift things up away from the earth, we are pulling against gravity
4   all other things have gravity too (*e.g.* the moon and sun causing tides).

*Note*

It is misleading to say 'gravity pulls'. It is the earth which pulls. *Gravity* is the name which we give to that pull. To say that gravity pulls is like saying that a pull pulls.

It has been said that the question 'Why do things fall down and never up?' is asked only by little children and great philosophers. When we say 'down' we mean towards the earth, and when we say 'up' we mean away from it, so that the question is really concerned with why things fall towards the earth.

The question 'Why do things fall towards the earth?' can be answered simply, by the statement that the earth attracts all things towards it. The reason for the earth's attraction is not yet fully understood. Nevertheless, the earth does force things to move towards it. This force of attraction is called gravity, and it will be noted that in the Pupils' Book, it is the pull of the earth which is referred to, and not the pull of gravity. The phrase 'pull of gravity' is misleading, as of course it is the earth which attracts, and not gravity.

The earth attracts things from all directions towards the centre of its gravitational field. This is why it is not possible to fall off the other side of the world. It is also why, if it were possible to climb down an imaginary shaft from Britain to Australia, you would not still be going 'down' when you came out in Australia. Once the centre of gravity of the earth were passed, then the direction would be away from it, *i.e.* upwards. But for gravity, we should all be flung off into space, owing to the spinning movement of the earth, whose speed of rotation at the equator is over 1600 kilometres per hour.

Because the earth pulls all things, everything will fall towards the lowest position on the earth, unless prevented by something underneath it. Things which float on a liquid, *e.g.* corks on water, only do so while the liquid is in the way. Similarly, things which are lighter than air, and which therefore 'float' in the air, would fall straight to the ground, if the air could be suddenly taken away. The tiny drops of water which make up a cloud, for example, are lighter than the air

115

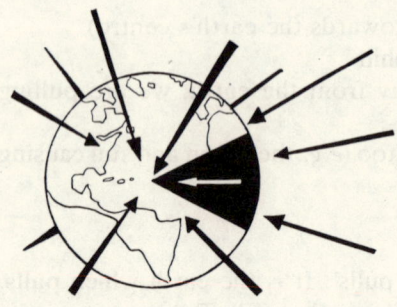

The earth attracts things from all
directions to its centre of gravity

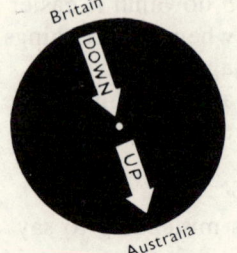

SECTION THROUGH THE EARTH

Towards the earth's centre of
gravity is 'DOWN'

Away from the earth's centre
of gravity is 'UP'

which supports them. Once they become bigger and heavy enough, they fall to the ground as rain. A feather and a parachutist would fall like stones if there were no air in the way.

As the force of attraction increases with nearness to the earth, a falling body will increase in speed as it approaches the earth. However, the earth is not an exact sphere, so that its pull is not equal everywhere. In consequence of this, the rate of acceleration varies slightly in different parts of the world.

An object falling through a liquid or a gas, such as water or air, obviously meets with resistance. The greater the surface area which the falling object presents, compared with its weight, the greater the resistance it meets. The greater the resistance it meets, the more it tends to be slowed down. For example, a large sheet of paper will not fall as swiftly as it would if it were screwed up into a ball. However if no liquid or gas were in the way (*i.e.* in a vacuum) the rate of fall would be the same. A parachute provides a falling weight with a large surface area and limits the speed of a falling man to about 19 kilometres per hour.

It is because of gravity that it is more difficult to force things to move uphill, and easier to force them to move downhill. Children who find a peculiar delight in rolling down a grassy slope are allowing gravity to force them to move downhill. However, although carrying, sliding and rolling may be easier downhill, there are times when it is necessary to resist gravity in order to prevent losing control. A hiker for example, carrying a rucksack down a steep slope, may encounter difficulties, and it may be necessary to apply brakes to a wheeled vehicle.

116

It is because of gravity that liquids and gases flow towards the lowest point they can reach on the earth. Liquids and gases, of course, have no shape to stop them flowing, and the heaviest naturally settles at the bottom. When air is heated, it expands and becomes lighter. As it becomes lighter, it is forced to move upwards. This is because gravity forces the surrounding colder and heavier air to flow in and replace it. If it were not for gravity, there would be no wind.

Gravity is responsible for the heaviest things sinking and the lightest being forced to rise. Consequently, if it were not for gravity, water vapour would not be able to leave the surface. There would be no air currents forcing clouds of condensed vapour across the land. There would be no rain falling from the skies, and no water flowing down to the seas.

So it is, that although moving air and moving water force other things to move from place to place, they in turn are forced to move by gravity. Of the three weather makers – sun, air and water – only air and water make their contributions because of gravity.

The gravitational field of the earth extends in all directions. However, as the distance from the earth increases, the effect of the earth's attraction becomes less. But although the effect would eventually be negligible, it is wrong to state that there is a point where the gravitational field of the earth ceases completely.

Gravity is one of the two major forces which may hinder living animals in their movement from one place to another. The other is friction. Consequently, when living animals move from place to place, or move other things from place to place, they have to overcome these two forces. Owing to the support provided by liquids, animals which live in water are not encumbered by the effects of gravity to the same extent as those which live on the land. When we throw a ball into the air, lift a pen off the desk, or send up an artificial satellite to circle the earth, we overcome the force of gravity, but only temporarily.

On the other hand, when we send a space vehicle towards the sun, or outwards towards the rim of the solar system, we need to be able to overcome the gravity of the earth permanently. In fact, the advent of space travel has at last belied the statement that 'What goes up must come down'.

## All Other Things Have Gravity Too

So far, it is only the force of attraction of the earth which has been

117

considered. However, every particle of matter in the universe has gravity and attracts every other particle. This means that although the earth pulls all things – alive, dead, or never-alive – towards it, they in turn pull the earth towards them. Not only that, but they exert a pull on each other. The force of attraction which a particle exerts, is relative to its mass. Consequently, although a grain of sand pulls the earth, its force of attraction is negligible compared with that of the earth, because its mass is so much less than that of the earth. In fact, most of the things with which we are familiar are so small in mass compared with the earth, that their force of gravity is looked upon as being ineffective. Nevertheless, an interesting demonstration carried out on the Scottish mountain Schiehallion showed that a mass such as a mountain could affect a plumb-bob sufficiently to move the plumbline out of the vertical.

Large masses such as the sun, moon and planets have, of course, considerable forces of gravity. It is the mutual force of attraction between the sun and its orbiting planets which prevents them from flying off into space. In the same way the mutual attraction of the earth and the moon prevents the moon from shooting off at a tangent. The pull of the sun and the pull of the moon are responsible for the rise and fall of the tides on earth.

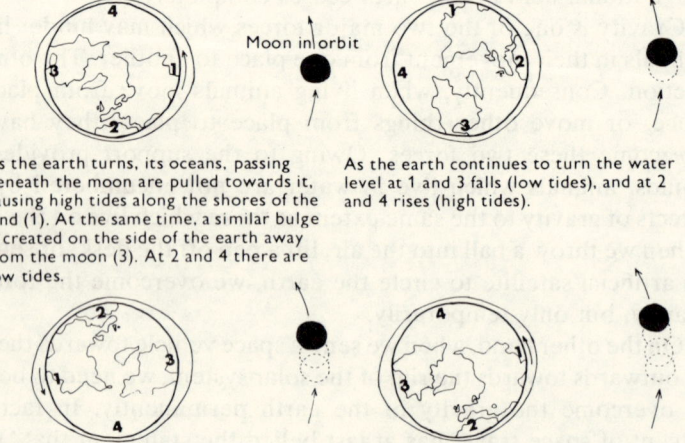

Moon in orbit

As the earth turns, its oceans, passing beneath the moon are pulled towards it, causing high tides along the shores of the land (1). At the same time, a similar bulge is created on the side of the earth away from the moon (3). At 2 and 4 there are low tides.

As the earth continues to turn the water level at 1 and 3 falls (low tides), and at 2 and 4 rises (high tides).

Half a day after its first high tide 3 is nearest to the moon and experiences a second high tide. There is a high tide at 1 also, and low tides at 2 and 4.

The earth has rotated once in a 24-hour period, during which time there have been two high tides and two low tides at 1, 2, 3 and 4. However, during this period the moon has moved on in its orbit so that the high tide now at 1 is an hour later than on the previous day.

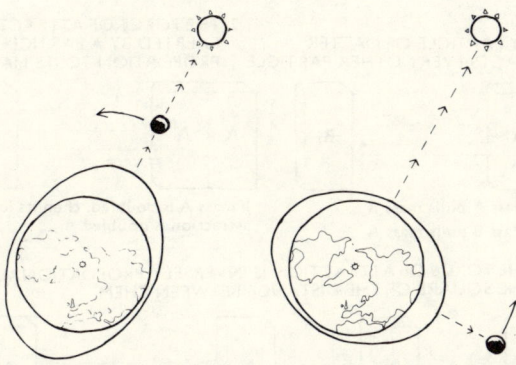

At New Moon and at Full Moon, the sun, earth and moon are in line so the pull of the moon and sun is in line, resulting in higher tides than average, *ie* Spring tides.

At Half Moon, the moon and sun are pulling at right angles to one another, resulting in lower tides than average, *ie* Neap tides.

THE PULL OF THE MOON AND SUN CAUSES DIFFERENT TIDES ON EARTH

## Some Technical Information

The force of attraction existing between any two particles is inversely proportional to the square of the distance between them. This means that the nearer two masses are to one another, then the stronger is the force of attraction between them. The further apart they are, the weaker is the force of attraction between them. Thus, although the sun is considerably more massive than the earth, its distance of some 150 million kilometres results in its effective pull on us being considerably less than that of the earth. Similarly, although the mass of the moon is far less than the mass of the sun, it is so much nearer to the earth, that its effect on the tides is greater than that of the sun. Forcing a space vehicle to move from the earth to the moon necessitates developing enough power to take the vehicle near enough the moon for its pull to be more effective than that of the earth. The same problem arises when sending a vehicle from the earth to another planet.

*Mass*, incidentally, has nothing to do with size. Mass may be loosely described as being the amount of matter in a body. Thus, if you had two balls of the same size, and composed of the same material, but one being solid right through and the other being hollow inside, then the hollow one would have the least mass and would consequently have the least pull. Similarly, a block of lead the size of a brick is denser, and has more mass than a block of sponge the size of a brick. Consequently, the block of lead would have a

119

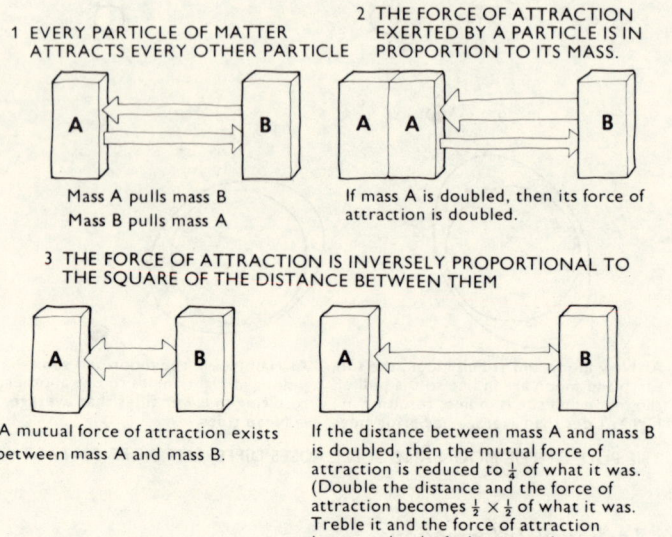

1 EVERY PARTICLE OF MATTER ATTRACTS EVERY OTHER PARTICLE

2 THE FORCE OF ATTRACTION EXERTED BY A PARTICLE IS IN PROPORTION TO ITS MASS.

Mass A pulls mass B
Mass B pulls mass A

If mass A is doubled, then its force of attraction is doubled.

3 THE FORCE OF ATTRACTION IS INVERSELY PROPORTIONAL TO THE SQUARE OF THE DISTANCE BETWEEN THEM

A mutual force of attraction exists between mass A and mass B.

If the distance between mass A and mass B is doubled, then the mutual force of attraction is reduced to $\frac{1}{4}$ of what it was. (Double the distance and the force of attraction becomes $\frac{1}{2} \times \frac{1}{2}$ of what it was. Treble it and the force of attraction becomes $\frac{1}{3} \times \frac{1}{3}$ of what it was.)

greater pull than the block of sponge. Summarising, the following may be said:

1 Every particle of matter attracts every other particle.
2 The force of attraction which a particle exerts, is in proportion to its mass.
3 The force of attraction existing between any two particles is inversely proportional to the square of the distance between them.

The word *gravitation* comes from the Latin, and means *heavy*. On the earth, what we mean by the 'weight' of a thing is really a measure of the force of attraction existing between the earth and that thing.

*Weight*, of course, is not the same as mass. The *mass* of a thing, or its amount of matter remains constant. Its weight however, being a measurement, varies according to the force of attraction existing between it and the earth. As the earth is not an exact sphere, its pull is not equal everywhere. That is to say, it is most effective at points on the surface nearest to its centre of gravity. Thus, as it is slightly flattened at the poles, and slightly bulging at the equator, it is most effective at the poles, and least effective at the equator. It is because of this, that the force of attraction between the earth and an object will, generally speaking, be greater at the poles, and less at the equator. Thus it is that a bag of sand which weighs 1 kilogram in the temperate zone will weigh a few grams more at the poles, and a few grams less at the equator. If the bag of sand is taken out into space, the force of attraction will decrease with distance, so that about 6000 kilometres from the surface of the earth its effective weight will be almost down to a quarter of a kilogram. Far enough out, the effective weight becomes negligible, so that an astronaut can leave a space vehicle without any fear of falling straight down to the earth. He simply drifts, weightless in space, if anything, gravitating slowly towards the nearest mass, *i.e.* the space vehicle itself.

If two objects were to be weighed at the same point relative to the earth, it would be true to say that the heavier of the two would be the one with the stronger pull. However, for practical purposes, it is obvious that things which children experience as being the heaviest are the ones with the strongest pull. Thus the desk is heavier than the pen, because its gravity is greater than that of the pen. Similarly, the gravity of the heaviest child in the classroom will be greater than the gravity of the lightest.

Although there had been some conception of gravity long before the time of Sir Isaac Newton, it was he who established the law of gravitation upon which mathematical calculations are based. It is said that the train of thought which led to the formulation of the law of gravitation was prompted by a falling apple in Newton's orchard at Woolsthorpe. The story that the apple fell on his head is an interesting one, but it should not be assumed that it is true.

## CODE

1 Demonstrate that because of gravity it is harder to force things to move uphill, and easier to force them to move downhill.
   A 1a Pull the brick by means of an elastic band and paper clip, so that it is forced to slide along a level surface.
     b Observe extent to which elastic band is stretched, while load is sliding.
    2a Tilt the surface and force the load to slide uphill.
     b Observe increased extent to which elastic band is stretched, while the load is sliding.
    3a Reverse direction, and force load to slide downhill.
     b Observe decreased extent to which elastic band is stretched, while the load is sliding.
   B Repeat A1 and A2, but this time substitute a suitably heavy toy vehicle on wheels. Alternatively, mount the brick on

121

rollers. No effort, of course, will be required to force the load to roll downhill.

2　Experiment to find the way towards the centre of the earth. Suspend a weight on a thread and swing it (*e.g.* a conker on a string). Observe that it always comes to rest pointing downwards. This is the principle of the plumb line used by builders to ascertain the vertical.

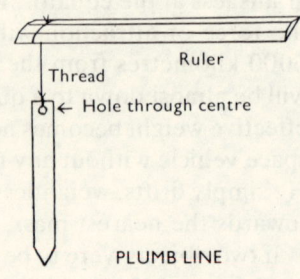

3　Observe that sports such as high jumping, long jumping, weight lifting, putting the shot, and throwing the discus or javelin are all tests of ability to overcome gravity.

4　Observe that the things that weigh the most at one particular place are actually the things with the greatest pull, *e.g.* a desk has more gravity than a book; a heavy child has more gravity than a light weight child.

**Written Work**

1　The earth has a <u>pull</u> that forces things to move down.
2　The pull of the <u>earth</u> is called <u>gravity</u>.
3　You pull against gravity when you <u>lift</u> things up.
4　The pull of the earth <u>forces</u> water to move.
5　Gravity makes it hard to carry <u>heavy</u> things.
6　Tides are caused by the pull of the <u>moon</u> and <u>sun</u>.

# QUESTIONS ON LESSONS 1 TO 12

1　What are the three kinds of things in the world?　*Alive, dead, never alive*

2　What are the two kinds of living things?　*Animals and plants*

3　What are the two kinds of dead things?　*Animal and plant*

4　What are the three forms in which we find never-alive things?　*Solid, liquid, gas*

| | | |
|---|---|---|
| 5 | Which of these three has both size and shape? | *Solid* |
| 6 | Which of these three has size but no shape? | *Liquid* |
| 7 | Which of these three has neither size nor shape? | *Gas* |
| 8 | Are living things in the form of solids, liquid or gas? | *Solid* |
| 9 | What is air a mixture of? | *Gases* |
| 10 | What are the four main needs of living things? | *Oxygen, food, to grow, to have young* |
| 11 | Which two of these do all living things need in order to grow bigger? | *Oxygen and food* |
| 12 | Of the plants which are found on land, what are the three kinds with roots, stems and leaves? | *Herbs, trees, shrubs* |
| 13 | Of these three, which have woody stems – the herbs, or the trees and shrubs? | *Trees and shrubs* |
| 14 | Which has one main stem growing from the roots – a tree or a shrub? | *A tree* |
| 15 | What name do we give to plants with no roots, stems or leaves? | *Simple plants* |
| 16 | Can any simple plant have flowers, fruits and seeds? | *No* |
| 17 | What are found inside fruits? | *Seeds* |
| 18 | Which plant part does a fruit grow out of? | *A flower* |
| 19 | Are eggs and seeds alive or dead? | *Alive* |
| 20 | What are the stages in the life of an insect with a three-stage life? | *Egg, larva, adult* |
| 21 | What are the stages in the life of an insect with a four-stage life? | *Egg, larva, pupa, adult* |
| 22 | Which is the only stage in which an insect can grow bigger? | *Larva(l)* |
| 23 | Which is the only stage in which an insect can have wings? | *Adult* |
| 24 | Is a caterpillar a larva or a pupa? | *A larva* |
| 25 | Is a whale a fish or a mammal? | *A mammal* |
| 26 | Is a bat a bird or a mammal? | *A mammal* |
| 27 | Which is the only one that can force itself to move from place to place – an animal, wind or water? | *An animal* |

28 What are the three main ways in which we force things to move from place to place on land? *Carrying, sliding, rolling*

29 Which is the easier path to slide things over – a rough path or a smooth path? *A smooth path*

30 Which is the easiest way to force a heavy round thing to move from place to place – carrying, sliding or rolling? *Rolling*

31 What do we call the force which makes it harder to carry something heavy than to slide it or roll it? *Gravity*

32 The three main things which force other things to move from place to place on land, are moving animals, moving liquid and moving . . .? *Gas (Air)*

33 What do we call the bar which joins two wheels together? *An axle*

34 What are the three weather makers? *Sun, air, water*

35 Which two of these help to dry wet things? *Sun and (moving) air*

36 What does water become when it dries up? *Water vapour*

37 Is water vapour a liquid or a gas? *A gas*

38 A cloud in the sky is made up of very tiny drops of what? *Water (Liquid)*

39 What forces the clouds to move from place to place? *Moving air (Wind)*

40 The force which causes things to fall towards the centre of the earth is called the force of . . . *Gravity*

**13**    # SOLIDS, LIQUIDS AND GASES

## MIXTURES OF THINGS

### Demonstration Material

1   Material suitable for illustrating a mixture of solids, *e.g.*
     *a* broken concrete, pebbles showing more than one colour
        sample of conglomerate rock, lump of ore
     *b* sand and salt.
2   Material suitable for illustrating a mixture of liquids, *e.g.*
     *a* milk
     *b* oil and water.
3   Material suitable for illustrating a mixture of solid and gas, *e.g.*
     *a* section of foam rubber, plastic sponge, or rubber ball, with air
        spaces clearly revealed
     *b* something suitable for making smoke, *e.g.* burning oily paper
        in a tin lid.

### Sample Link Questions

1   What are the three different forms in which never-alive things
     are found? (*Solid, liquid, gas*)
2   Which of these three forms has both size and shape? (*A solid*)
3   Which has size but no shape? (*A liquid*)
4   Which has neither size nor shape? (*A gas*)
5   What is air a mixture of? (*Gases*)
6   When water dries up, what does it become? (*Water vapour*)
7   Is water vapour solid, liquid or gas? (*Gas*)
8   What is a cloud made up of? (*Tiny drops of water*)
9   What name do we give to a cloud near the ground? (*Mist or fog*)

### Relevant Information

The main purpose of this lesson is to show:

1   many never-alive things are mixtures. Some mixtures have a
     solid form, some have a liquid form, some have the form of a gas
2   some mixtures are of solids, some of liquids, and some are of
     gases

3   solids, liquids and gases may be mixed with one another, *e.g.*
    *a* smoke is a mixture of gas and bits of solid
    *b* town fog is a mixture of solid, liquid and gas.

When two or more substances are mixed together there may be

  *a* a chemical reaction during which some other substance or
      substances will be formed;
*or b* no chemical reaction, in which case the result is simply a
      mixture of the substances used.

It is with the second possibility that this and the following lesson are
concerned.

In a *mixture* the constituents are capable of separation by physical
means. Except in the case of certain solutions, these constituents
can be mixed in any proportion. There may be any number of these
constituents, but so long as none of them reacts chemically with any
of the others, then they remain a mixture. The mixture itself can be
solid, liquid or gas in form, depending upon its constituents. The
following are the different kinds of simple mixtures which can be
obtained from solids, liquids and gases.

1   A mixture of solids
2   A mixture of liquids
3   A mixture of gases
4   A mixture of solid and gas
5   A mixture of solid, liquid and gas
6   A mixture of liquid and solid
7   A mixture of liquid and gas

In some cases when two substances are mixed together, they remain
just as they are. For example, if a box of drawing pins were emptied
into a jar of oil, then the jar would contain a mixture of oil and
drawing pins.

In other cases when two substances are mixed together, the result
may be that one subdivides into smaller particles which disperse
throughout the other. Such mixtures are of two kinds:

1   Colloidal mixtures          2   Solutions

*Colloidal Mixtures.* In a colloidal mixture, a solid, liquid or gas is
finely subdivided into minute particles which are suspended in some
other solid, liquid or gas, but they are not the smallest particles
possible.

*Solutions*. A solution is a mixture in which a solid, liquid or gas is finely subdivided into the smallest possible particles (molecules), which are dispersed throughout some other solid, liquid or gas. Solutions are referred to as *molecular mixtures*.

Lesson 13 in the Pupils' Book is concerned with mixtures in general, and principally with the first five kinds of mixtures listed above. It is not concerned with the difference between a colloidal mixture and a solution. Lesson 14 in the Pupils' Book is concerned with the last two kinds of mixtures, namely mixtures of liquid and solid, and mixtures of liquid and gas, and also with the term *dissolve*.

## 1   *Mixtures of Solids*

A mixture of solids is usually easy to recognise as such. A box full of an assortment of nuts, bolts and screws, is a mixture of solids. A mixture of sand and salt is a mixture of solids. A piece of broken concrete should reveal signs of the different solids used when the concrete was mixed. Mixtures of rocks are found in samples of millstone grit, 'pudding stone', and other conglomerates. Rock itself is a mixture of the minerals of which the crust of the earth is composed. In many pebbles different solids are easily distinguished by their colour.

## 2   *Mixtures of Liquids*

When some liquids are mixed together, one dissolves in the other, forming a solution. This happens when an alcohol (*e.g.* methylated spirit) is mixed with water. When other liquids are mixed together, the particles of one will become dispersed throughout the other. Such a mixture of liquids is known as an *emulsion*. In general, emulsions are oil and water mixtures and are of two main kinds, according to which is dispersed in which.

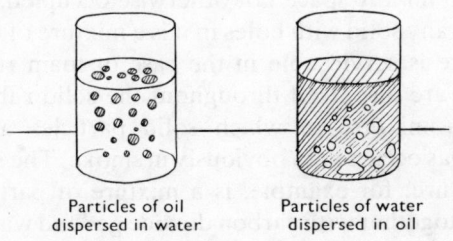

Particles of oil
dispersed in water

Particles of water
dispersed in oil

Normally when oil and water are mixed together, they will not remain as an emulsion, but will separate, the heavier water sinking to the bottom, and the oil being forced to float on top. However, the mixture can remain as an emulsion if there are other substances present to prevent the dispersed particles from collecting together. Such substances are known as *emulsifying agents*. Mayonnaise is basically an emulsion of vinegar and olive oil, in which egg yolk is mixed to prevent the two liquids from separating. Other examples of emulsions are salad dressing, and various ointments and creams. Margarine is an emulsion of water in oil, stabilised by a small percentage of other substances. Medicinal cod liver oil is an emulsion consisting basically of oil from cod's liver, mixed in water and stabilised by a small amount of other substances.

The fact that oil is mixed with another liquid does not prevent it from solidifying, *i.e.* becoming fat, if there is a sufficient drop in temperature. For example, fresh milk is, at the body temperature of a mammal, an emulsion of oil and water, in which other substances are present. But as the milk cools, the oil particles solidify into fat particles, so that milk is not always an emulsion of oil and water.

## 3   Mixtures of Gases

Gases are rarely encountered on their own, as once they escape into the air, they mix freely with the other gases there. The gases of the air really form a molecular mixture; the two chief gases are nitrogen and oxygen. The amount of water vapour present varies from place to place, and there are traces of other gases such as carbon dioxide.

Coal gas (sometimes called just 'gas') is itself a mixture of gases. It consists mainly of methane, carbon monoxide and hydrogen, with traces of nitrogen, carbon dioxide and oxygen.

## 4   Mixtures of Solid and Gas

As air tends to fill any space not otherwise occupied, it is perhaps fair to say that any solid with holes in it is a mixture of solid and gas. Such a mixture is serviceable in the case of foam rubber, where pockets of air are dispersed throughout the solid rubber.

However, a mixture in which solid particles are dispersed throughout a gas occurs very obviously in smoke. The smoke from a wood or coal fire, for example, is a mixture of particles of solid carbon (soot) together with carbon dioxide gas and water vapour. It

128

is the solid unburnt particles which make a smoke visible. American Indians developed a system of signalling with puffs of smoke. Smoke has also been used in warfare for screening purposes. The solid particles in domestic and industrial smokes have been responsible for damage to health, property and vegetation, and for reducing the value of sunlight. This is the reason for proclaiming smokeless zones in city areas.

### 5 Mixtures of Solid, Liquid and Gas

Many things may be a mixture of solid, liquid and gas, without readily being apparent as such. Fog is the example used in the Pupils' Book as being easily experienced as such. The thick fog of cities is caused by water vapour in the air condensing in droplets on the tiny particles of carbon and other solids present in a smoky atmosphere.

## CODE

### 1 Mixtures of Solids

a Examine broken concrete, broken pebbles, ores, or samples of conglomerate rocks for mixtures of solids.
b Make a mixture of solids, e.g. with sand and salt.

### 2 Mixtures of Liquids

a Shake a bottle of milk, so that the cream mixes with the remainder of the milk.
Observe how cream reassembles at the top. (This is, of course, due to gravity – the heaviest substance sinking to the bottom and the lightest being forced to the top.)
Observe the appearance of a watery mixture at the bottom of a bottle of sour milk.
b Make an emulsion of two liquids, e.g. water and machine oil, water and paraffin, water and engine oil.
Observe how the heavier liquid – in each case water – eventually settles to the bottom.

### 3 Mixtures of Solid and Gas

a Observe air spaces in a piece of foam rubber, plastic sponge or

129

rubber ball. A piece of sponge may be squeezed under water to release the air bubbles.

*b* Make a smoke, *e.g.* light a piece of crumpled oily paper in a tin lid, and blow gently on the burning section.

## 4 *Mixture of Solid, Liquid and Gas*

Observe particles of dust and water which collect on clothing or vehicles, etc., in fog.

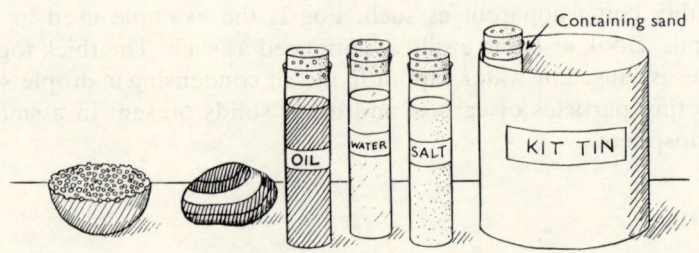

Part of rubber ball, pebble, tubes, and tin container

*Note*

Capped specimen tubes are useful both for containing and mixing those solids and liquids which are to be used by children working individually or in groups. Tins such as coffee tins are suitable for storing separate sets of equipment for this lesson.

## Written Work

1 Many rocks are a mixture of solids.
2 Oily water is a mixture of liquids.
3 Coal gas is a mixture of gases.
4 Smoke is a mixture of gases and solid.
5 Fog is a mixture of solid, liquid and gas.

# 14

# MIXING THINGS

## DIFFERENT KINDS OF MIXTURES

### Demonstration Material

Any or all of the following:

1  Any solid suitable for making a paste in water, *e.g.* school paste powder, clay powder, flour, starch.
2  Soap and water or detergent and water for making a foam.
3  Any solid which will dissolve in water, *e.g.* potassium permanganate, ink powder, sugar, etc.
4  Top half of a plastic detergent bottle, paper tissues, glass jar.
5  Two liquids which will make a solution of liquids when mixed together, *e.g.* Dettol and water, methylated spirit and water, oil and turpentine, oil and paraffin, turpentine and paraffin.
6  White card or paper, jam jars, capped specimen tubes, water.

### Sample Link Questions

1  Is coal gas just one gas or a mixture of gases? (*A mixture of gases*)
2  What is the commonest of all mixtures of gases? (*Air*)
3  What are most rocks a mixture of? (*Solids*)
4  What kind of mixture does oil and water make? (*A mixture of liquids*)
5  What is smoke a mixture of? (*Solid and gas*)
6  What part of the smoke do we see? (*The solid part*)
7  What is fog a mixture of? (*Solid, liquid and gas*)
8  What things make up a fog? (*Smoke and tiny drops of water*)
9  What makes a fog so hard to see through? (*The tiny drops of water and bits of solid*)

### Relevant Information

The main purpose of this lesson is to show:

1  some mixtures are of liquid and solid
2  some mixtures are of liquid and gas
3  a solution is a mixture.

The two kinds of mixtures not dealt with in Lesson 13 are mixtures of liquid and solid, and mixtures of liquid and gas.

## 5  Mixtures of Liquid and Solid

A colloidal mixture of liquid and solid is known as a *paste* and results from the concentrated dispersal of fine solid particles throughout a liquid. A paste may be plastic and easily moulded; but whether it flows like a liquid, or retains its shape like a solid, depends mainly on the size, shape, and concentration of the solid particles. Examples of pastes are flour and water mixes, school and household pastes, toothpaste, mud, paints, potter's clay, and putty. Putty is a paste of whiting (calcium carbonate) and linseed oil.

## 6  Mixtures of Liquid and Gas

A colloidal mixture of liquid and gas may have particles of gas dispersed throughout the liquid, or particles of liquid dispersed throughout the gas.

When a mixture of liquid and gas consists of particles of gas dispersed throughout the liquid, the mixture is known as a *foam*. In foams the gaseous particles are usually air. Pure liquids do not permit good, lasting foams. The addition of a third component which acts as a foaming agent, allows the formation of a foam which is more stable; soap in water, for example, promotes a better foam than would the water on its own. Typical foams are found on breaking waves, at the foot of a waterfall, and on soapy water. Foam does, of course, occur on liquids other than water and is not infrequently referred to as 'froth'.

When a mixture of liquid and gas consists of particles of liquid dispersed in a gas, then the mixture comes into the category of an *aerosol*. A cloud in the sky, or the cloud of condensed steam from a kettle, consists of particles of liquid dispersed in the air.

### Dissolving

When two substances are mixed together, and one of them subdivides very finely into its smallest possible particles (molecules), it is said to have dissolved. The resulting mixture is then known as a *solution*. In a solution, the molecules of a solid, liquid or gas are dispersed throughout some other solid, liquid or gas, and the result is a molecular mixture. The dispersed substance is known as the *solute*, and the substance throughout which it is dispersed, is known as the *solvent*.

It is possible to have a solution where the solvent is a solid. Hydrogen gas is absorbed by certain metals which can then be quite fairly known as solid solutions. Antimony and bismuth form a solid solution. It is also possible to have a solution where a gas is the solvent. For that matter, the gases in the air form a molecular mixture, and could therefore be strictly regarded as being in solution. The most common solutions, however, have a liquid as the solvent.

## Liquids as Solvents

Water is the most common solvent, and most inorganic substances will dissolve in it, at least to a certain extent. Even rock, glass and metals dissolve in it slightly, and pure distilled water is difficult to maintain as such. Sea water is a solution containing a higher percentage of salt, together with smaller quantities of other substances. Some substances, of course, dissolve more easily in other liquids than they do in water. Oil and cooking fat dissolve easily in petrol. Sulphur, oil and camphor dissolve in turpentine. Oils and fats will dissolve in ether.

## Substances Dissolving in Liquids

a *Solids*. The amount of a particular solid which will dissolve in a particular liquid varies with the temperature of that liquid. For most solids, this means that more can be dissolved in, say, hot water than in cold water. When the quantity of solid dispersing throughout a liquid has reached its maximum, *i.e.* when no more will dissolve in it, then the solution is said to be *saturated*.
When a solid is dissolved in a liquid, the particles of solid – being molecules – are so small that they will not be separated from the liquid when the solution is passed through a filter.

b *Liquids*. When two liquids form a molecular mixture, either may be considered as dissolving in the other. For example, if a thimbleful of alcohol is mixed with a bucketful of water, then the molecules of the alcohol will be dispersed throughout the water, and it may be said that the alcohol has dissolved in the water. On the other hand, if a thimbleful of water is mixed with a bucketful of

133

alcohol, then the molecules of water will be dispersed throughout the alcohol, and it may be said that the water has dissolved in the alcohol. Obviously if the mixture consists of equal amounts of each liquid, the solution may be considered as one in which each liquid has dissolved in the other. Obviously two liquids which can exist in solution with one another may be mixed in any proportion.

c *Gases*. The amount of a particular gas which will dissolve in a particular liquid varies with the temperature of that liquid. Less gas will dissolve in, say, hot water than in cold water. This is because raising the temperature of a liquid tends to drive out the gases which are in solution. When the quantity of a gas dispersing throughout a liquid has reached its maximum, *i.e.* when no more of it will dissolve in the liquid, then the solution is said to be saturated for that gas.

The gases of the air are slightly soluble in water. It is oxygen from the air dissolved in water which is so important to animals and plants which absorb their oxygen from water.

Ammonia is the most soluble of all known gases in water. It is estimated that about 800 volumes of ammonia gas will dissolve in one volume of water at a temperature of 15°C.

## Notes

1 Dissolving is not the same as a chemical reaction. For example, a metal does not dissolve in an acid. The acid and the metal react together to form other substances.

2 Dissolving is not the same as melting. Melting occurs when, owing to a rise in temperature, a substance changes from the solid to the liquid state. For example, at temperatures above freezing point, ice melts and becomes water. It melts from the solid state into the liquid state. It does not dissolve.

## CODE

1 Experiment to make a *paste* from any of these liquids and solids:
   a school paste powder and water
   b clay powder and water
   c flour and water
   d starch and water
   e powdered paint and water

2 a Experiment to obtain a *foam* from liquid and gas, *e.g.* shake

up soapy water or detergent and water, and observe that the resultant foam is a mixture of bubbles of air and liquid.

*b* Observe other foams, *e.g.* that on a soft drink or mineral water, which is caused by carbon dioxide in flavoured water – the constituents of many such drinks.

3 Experiment to obtain a solution of a solid in a liquid, *e.g.*

*a* Drop particles of potassium permanganate into a jar of water and observe the gradual diffusion of colour throughout the water.

*b* Mix a very small amount of ink powder in water and observe the diffusion of colour.

*c* Observe that when something dissolves, the molecules are very tiny. Cut off the top half of a polythene detergent bottle, and line the inside with two paper tissues. (The half detergent bottle should be washed free of detergent, and the tissues should not be waterproof.)

Invert this funnel over a jar, and place some sugar inside the tissues, observing that the sugar does not pass through them. *Gently* add water to the sugar, and observe that water passes through the tissues into the jar.

Taste the dissolved sugar in the water in the jar, and observe that the molecules of sugar must have been very tiny in order to pass through the unseen holes in the paper tissues.

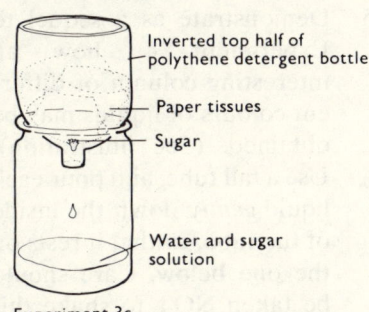

Inverted top half of polythene detergent bottle

Paper tissues

Sugar

Water and sugar solution

Experiment 3c

*d* Demonstrate the extensive dispersal of molecules in a solution:

Stir one particle of potassium permanganate into, say a jam jar of water until fully dissolved. Observe colour.

Pour half of this solution into a second jam jar which is already half filled with water. Observe presence of colour in diluted solution.

Pour half of this diluted solution into a third jar which is half filled with water. Again observe that colour is still present.

Repeat with a series of jam jars, until a stage of dilution is reached where no colour is apparent. Line up these jars against a white background and compare with a jar of clear water.

4   Experiment to obtain a *solution* of liquids, *e.g.* mix together small quantities of any two of the following liquids: water, Dettol, turpentine, methylated spirits, machine oil, paraffin, liquid detergent. Observe:

   *a* that those pairs of liquids which stay mixed form a solution, *e.g.*

   *i* methylated spirit and water
   *ii* oil and turpentine
   *iii* oil and paraffin
   *iv* turpentine and paraffin
   *v* Dettol and water

   *b* that any pairs of liquids which eventually separate do not form a solution.

5   Demonstrate as a sequel to Experiment 4, how an interesting column of different colours of liquids may be obtained. (See illustration). Use a tall tube, and pour each liquid *gently* down the inside of the tube, so that it rests on the one below. Care should be taken NOT to shake this tube. Once arranged, it should be placed in a position where it will not be disturbed.

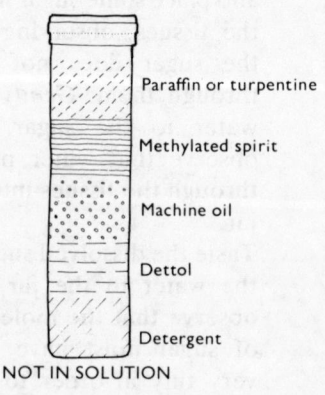

Paraffin or turpentine

Methylated spirit

Machine oil

Dettol

Detergent

NOT IN SOLUTION

*Note*

As in Lesson 13, capped specimen tubes are useful for shaking and mixing, when children are working individually or in groups.

**Written Work**

1   A paste is a <u>mixture</u> of liquid and <u>solid.</u>
2   A foam is a <u>mixture</u> of <u>liquid</u> and <u>gas.</u>
3   A solution is a mixture of tiny bits called <u>molecules.</u>
4   In most solutions, something is <u>dissolved</u> in a liquid.

# 15

# A LIQUID WE NEED

## WATER

### Demonstration Material

1    *a* Two glass jars, *e.g.* jam jars,
     *b* sand or aquarium gravel,
     *c* common clay or plasticine,
     *d* coloured water.
2    *a* Wide-bore glass or plastic tube,
     *b* straight-sided glass container, *e.g.* drinking tumbler,
     *c* aquarium gravel,
     *d* coloured water.
3    *a* Plant pot, or top half of plastic detergent bottle,
     *b* cotton wool,
     *c* sand,
     *d* jam jar
     *e* muddy water (mixture of soil and water).

### Sample Link Questions

1   What are the four main needs of living things? (*Oxygen, food, to grow, to have young*)
2   What is the liquid that all living things need for food? (*Water*)
3   What are the three weather makers? (*Sun, air, water*)
4   When water dries up, what does it become? (*Water vapour*)
5   Is water vapour solid, liquid or gas? (*Gas*)
6   Where does most of the water vapour in the air come from? (*The sea*)
7   When water vapour cools, what is formed? (*Clouds of tiny drops of water*)
8   What forces the clouds of tiny drops of water to move over the land? (*Moving air*) (*Wind*)
9   When these drops of water are heavy enough, what forces them to fall? (*Gravity*)
10   Gravity forces the waters of streams and rivers to move. Towards what? (*The sea*)

## Relevant Information

In Pupils' Book 1, water was introduced as being a food which all living plants and animals need.

The main points of this lesson are:

1   some animals obtain all the water they need from other foods;
2   other animals – including ourselves – need extra quantities, and take it by drinking it;
3   we obtain our drinking water in two main ways:
    a  from underground sources
    b  from surface sources.

Life, as we know it on this planet, began in water and still depends on water. It is a natural resource essential to the existence of all living things. It furnishes both plants and animals with food. It also acts as a solvent for other foods which may thus be absorbed into the various parts of the body. In addition, it is the basis of sap in those plants which have sap, and blood in those animals which have blood. No growth is possible without it.

Simple plants – the ones without roots, stems and leaves – absorb water containing dissolved foods directly through their surfaces. Herbs, trees and shrubs absorb water containing dissolved foods through the root hairs on their roots.

The water present in their normal food is enough for some animals. Other animals – including human beings – need more water than they can acquire in this way. They obtain more by drinking it.

All the water which living things use comes from the sea. Those which live in the sea obtain it directly; those which live on land obtain it indirectly. Water itself is perpetually circulating, from the sea to the air, from the air to the land, and from the land back to the sea. The stages are briefly:

1   Assisted by the sun and moving air, water from the surface of the sea becomes water vapour, which, being a gas, mixes with the other gases in the air.
2   When the air cools, water vapour in it condenses into tiny drops of water, forming clouds.
3   Moving air – wind – carries the clouds overland, where further cooling results in the formation of larger drops. Owing to the force of gravity, these larger drops fall to earth.
4   Again owing to the force of gravity, this water flows over the

ground and through the ground, reappearing as springs, which supply the streams and rivers.

5   Still owing to the force of gravity, these streams and rivers flow down to the sea. Water from the surface of the sea becomes water vapour, and so the cycle goes on. In some cases water may flow down to inland ponds and lakes from the surfaces of which it is evaporated in the same way as from the surface of the sea.

Animals and plants which live on land absorb and utilise some of the water which falls on the land, retaining a quantity of it for the time they are alive. When they die, however, they lose this water, which once again joins the never-ending circulation of water through the air and earth.

Human beings are animals requiring considerable quantities of water. Something like three-quarters of the weight of the human body is water, and one of the major problems of civilisation is how to obtain it. For although a human being may be able to survive for weeks without food, he can live only for a short time without water. Like many of the higher animals, we take in water in two main ways.

## 1   *By Consuming Foods Containing Water*

Both animal and plant foods have water in them. Plant foods – especially those such as potatoes and fruit – have a high water content. A hen's egg is over 73% water.

## 2   *By Drinking the Water Directly as a Liquid*

We may not drink much plain water, but it is present in other liquids that we drink, *e.g.* tea, coffee, cocoa, wines and beer. Mineral waters consist largely of flavoured water, and even cow's milk is over four-fifths water.

Our fresh water comes from two general sources:

1   from water lying in porous underground rocks
2   from water at the surface of the earth, *e.g.* in a stream, river, lake, waterhole, oasis

### *Water Obtained from Below the Surface – Ground Water*

Some rocks are not solid throughout, but contain spaces which vary in size from minute pores to large caverns. Water will percolate

139

through such porous rocks, just as it percolates through soil. Sandstone, millstone grit, chalk and limestone are examples of rocks which water can penetrate in this way.

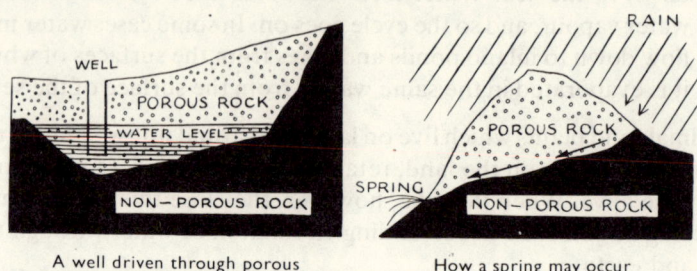

A well driven through porous rock into the section where the rock is saturated

How a spring may occur

When water, sinking through porous rocks, comes up against a layer of non-porous rock such as clay or slate, it can penetrate no lower. It therefore accumulates, saturating the porous rock until a level is reached where some of it can escape, possibly emerging at that point as a spring. This level is known as the water table.

A well sunk into the ground must reach the saturated zone below the water table. Water will seep into the bottom of the well from the saturated rocks which surround it, and as this water is either drawn or pumped out, so more will seep in to replace it.

*Note*

In the Pupils' Book a sectional view of a pump is shown. This is merely to illustrate that the water comes from spaces in underground rocks, and is not intended to show the mechanism of a pump.

The simple method of raising water from a well was by bucket and winch. During the nineteenth century, steam-driven beam pumping engines were used, and some of these are still in service and working well. Modern water pumps are often of the high speed rotary type.

Many farms, villages and small towns obtain their water from wells, and so do some large cities. Berlin's water supply, for example, comes almost entirely from wells. Altogether, it is estimated that nearly half the population of the world is supplied by water from wells.

Water from deep wells is to be preferred to water from shallow

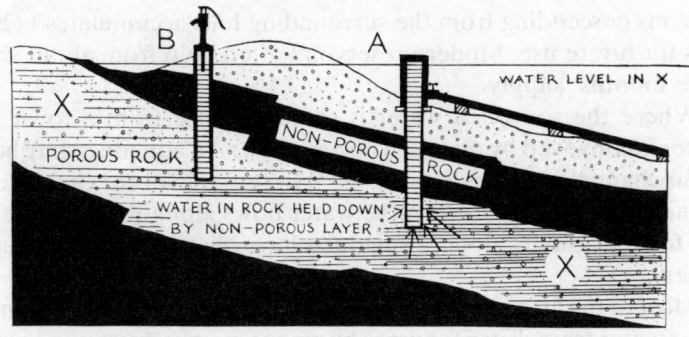

A     Artesian Well

If a flowing well is required, a shaft has to be driven through the non-porous (impervious) rock to release the artesian water which is held in the saturated porous rock below.

The point of outflow of the well must be below the water level of the saturated porous rock, so that the water will rise in the well shaft and flow out. A fair quantity of London's water comes from artesian wells, some of which are over 90 metres deep

B     Sub-artesian Well

If the point of outlet of the well is above the water level of the porous layer, the well is a non-flowing or sub-artesian well. Pumping is required to raise the water.

wells. Water at or near the surface tends to contain impurities, whereas the deeper water percolates, the more likely it is that these impurities will be filtered out. Living things occupy fresh water at the surface, including bacteria plants, amongst which may be the germs of various diseases. A shallow well near a farmyard is likely to contain unhealthy water for this reason. The great plague in London in 1665 was due to poor sanitation, the germs in the refuse in the streets often being washed straight into shallow wells and streams from which drinking water was obtained.

*Water Obtained from the Surface – Surface Water*

Water from below the surface of the earth emerges at the surface in the form of springs, from which streams and rivers result. Water for drinking purposes may be obtained direct from such rivers. London, for example, pumps water from the River Thames and its tributary the River Lea.

    Natural lakes serve as another source. Where the water supply is inadequate, an artificial lake – or reservoir – may be created. The principle of the reservoir is simple. It involves building a wall – known as a dam – across a valley, so that water from one or more streams descending from the surrounding hills accumulates behind this for future use. Modern reservoirs often hold from about six to nine months' supply.

Where the source of water is some distance from a town, the water is conveyed by means of an *aqueduct*. An aqueduct may be an open channel or a closed conduit. Where the inlet for the water is higher than the outlet, then the water flows along the aqueduct by the force of gravity alone. Aqueducts have been in use since ancient times.

Most of the large towns and cities of Great Britain obtain some of their water from distant sources by means of aqueducts, the notable exception being London. Glasgow obtains water from Loch Katrine; Liverpool obtains it from Wales. Manchester obtains water from Haweswater and Thirlmere, in the Lake District, by the longest aqueducts in the country. The water is piped over 160 km.

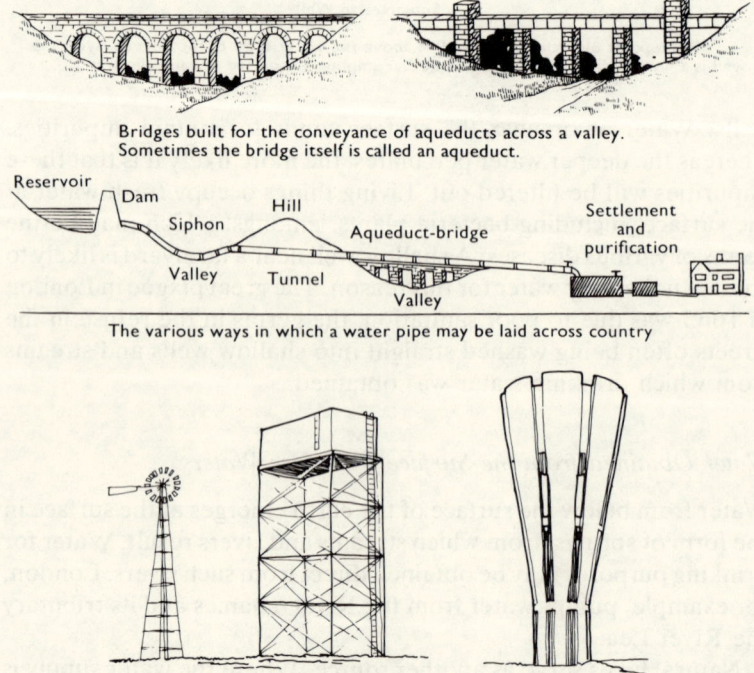

Bridges built for the conveyance of aqueducts across a valley. Sometimes the bridge itself is called an aqueduct.

The various ways in which a water pipe may be laid across country

WATER TOWERS

Where water from surface sources is not higher than the region it is to supply, it has to be pumped up to the required level. In some cases it is pumped to the top of specially constructed water towers, from which it can flow downwards into the water tanks of houses.

Water may at first be free from impurities, but as it descends on its way towards the sea, it can collect many. In consequence, water which is to be used for domestic purposes is purified before it reaches the consumer. This involves killing unwanted living things – especially harmful bacteria – and removing suspended particles. Larger particles are allowed to settle out in a settling tank; smaller lighter particles are filtered out through a filtration tank. If the water requires sterilising, chlorine gas or even ozone is used. Finally the water is stored.

Human beings, of course, use water for many purposes other than drinking, particularly for sanitation, irrigation, and for the production of power, so that in a city the size of London, the consumption may total something in the region of 1500 million litres per day.

## CODE

1 Demonstrate that some rocks have spaces in which water will lodge, whereas other rocks have not.

   a Half fill one glass jar with sand or aquarium gravel.

   b Half fill another glass jar with common clay or plasticine, well packed down.

   c Pour the same quantity of water into each, and observe how the water soaks into the sand or gravel but not into the clay.

2 Demonstrate how water soaks into the bottom of a well from the spaces in the rocks around it.

   a Stand a length of glass or plastic tubing (bore at least 13 mm) against the inside of a straight-sided glass container, e.g. a drinking tumbler, and fasten it to the tumbler with a strip of wide cellulose tape, running right down the tube.

   b Fill the tumbler with aquarium gravel.

   c Pour in coloured water, and observe how the water flows into the 'well'.

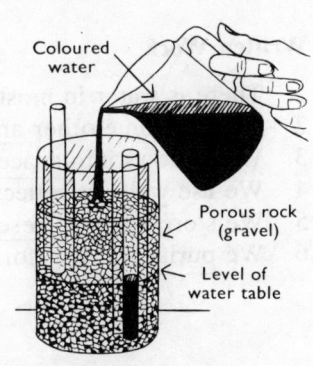

Coloured water

Porous rock (gravel)

Level of water table

The base of the tumbler acts as non-porous rock

*Note*

A second 'well', positioned at a higher level, will show the necessity for a well to be driven below the level of the water table.

3   Demonstrate how the solid particles in unclean water can be removed in a filter bed.

    *a* Plug the hole in a clay plant pot with cotton wool, and half fill the pot with sand.

    *b* Stand the plant pot on a glass container for collecting the filtered water.

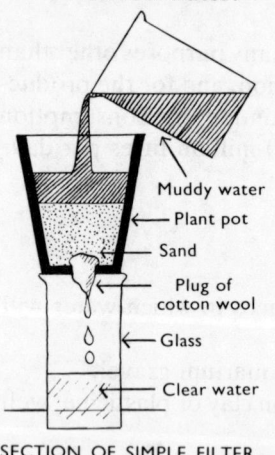

    *c* Mix some soil and water in a separate jar, and pour the mixture gently on to the sand. A small tin lid placed on the surface of the sand will prevent undue disturbance.

    *d* Observe how the solid particles are filtered out of the mixture by the sand, and clear water drips into the collecting vessel.

Muddy water

Plant pot

Sand

Plug of cotton wool

Glass

Clear water

SECTION OF SIMPLE FILTER

*Note*

The top half of a plastic detergent bottle may be used instead of the plant pot.

4   At the public baths, notice the smell of chlorine gas used for purification purposes.

## Written Work

1   There is <u>water</u> in most of our foods.
2   We and some other <u>animals</u> drink water.
3   Water flows into <u>spaces</u> in the rocks.
4   We use <u>wells</u> to collect water from below the ground.
5   We store water in <u>reservoirs.</u>
6   We <u>purify</u> water before we drink it.

144

# 16

# WATER IN WINTER

## FREEZING AND MELTING

### Demonstration Material

As many as possible of the following:
1 Lumps of ice in a jar of water.
2 Snow.
3 Material for making frost –
   *a* a tall tin,
   *b* crushed ice,
   *c* common salt.

### Sample Link Questions

1   Which is the coldest season of the year? (*Winter*)
2   What are the three weather makers? (*Sun, air, water*)
3   Which two things help water to become water vapour? (*Sun and air*)
4   What causes water vapour to change into tiny drops of liquid? (*Cooling*)
5   What are the three forms of never-alive things? (*Solid, liquid, gas*)
6   What is air a mixture of? (*Gases*)
7   Name two gases which you know are in the air. (*Oxygen, water vapour*)
8   Are snow, ice and hail solid forms or liquid forms? (*Solid*)
9   Does ice float on water or sink? (*It floats*)
10  What forces rain to fall down and water to flow? (*Gravity*)

### Relevant Information

The main point of a lesson in Pupils' Book 1 was that ice, hail, snow and frost are frozen forms of water.

   The main points of this lesson are:
1   freezing means changing to the solid form;
2   melting means changing from the solid to the liquid form (as distinct from dissolving which is a way of mixing);

145

3  the solid forms of water differ because they are formed in different ways.

*a* Ice and hail are caused by water freezing.

*b* Snow is caused by water vapour freezing.

When any liquid changes to the solid form, it is said to freeze. Freezing is brought about by cooling. In conditions of normal atmospheric pressure (*i.e.* 760 mm of mercury), water will freeze when the temperature is lowered to 0°Celsius.

Water is in contact with land in all of its three forms.

1  *As a gas*. Air is in contact with land, and water vapour is one of the gases in air.

2  *As a liquid* – the form in which it is most commonly experienced

3  *As a solid* – as ice, frost and snow, where conditions are sufficiently cold.

Water is also present in the air in all of its three forms.

1  *As a gas*. Normally it is present only in very small amounts. (It has been estimated that the water vapour content varies from 0.01% to 4%.)

2  *As a liquid* – the tiny drops of water which form clouds.

3  *As a solid*. The highest of the cirrus clouds, *e.g.* 'mares' tails', are composed of tiny particles of ice.

Not only liquid water may become solid; water vapour in the air itself will crystallise into solid particles if the temperature is sufficiently lowered. Whether the solid form presents the appearance of ice, hail, frost or snow depends upon the circumstances under which freezing takes place.

*Snow*

Snow is formed by the water vapour in the air condensing at a temperature below freezing point, so that, instead of forming droplets of liquid, it is precipitated as crystals which combine to form snowflakes. Snowflakes are therefore made up of tiny ice crystals, and these vary greatly in design and pattern, so that no two can be found alike. Their shape, however, is almost always a hexagon.

The size of flakes varies from that of the tiny 'diamond dust' crystals, which have a diameter of about .127 mm, to that of large cottony flakes, which may have a diameter of several centimetres and contain hundreds of individual crystals.

In the British Isles snow is rare at Christmas. It occurs chiefly during the latter part of winter and in early spring.

Damp snow makes a better snowball than dry snow, owing to the phenomenon known as *regelation*, which means 'freezing again'. Snow on the point of melting is damp, and, when compressed, the surfaces of the ice crystals which form the flake begin to melt and then freeze together, thus making a more compact snowball.

Being a bad conductor of heat, a blanket of snow can have the same effect as a blanket of wool and serve as a protection against loss of heat for any plants or animals which it covers. It is for this reason that plants covered by snow are protected from damage by frost. In the same way an igloo protects its occupants against excessive loss of heat.

The whiteness of snow is due to its reflecting something like 78% of the light which falls upon it and scattering it, so that it appears white when seen from any direction. This also accounts for the whiteness of foam.

## Frost

Frost occurs when the temperature of a surface in contact with the air falls below freezing point. It may be due to water vapour from the air changing into ice crystals on the cold surface, or it may be due to the freezing of drops of water which are already on the surface. Dew, for example, may freeze on the ground or on plants, when the temperature drops below freezing point. Frost occurs on the inside of a window pane, if the temperature of the pane drops below freezing point. Again, frost on a pane may result from the precipitation of ice crystals direct from water vapour in the room, or from liquid condensation which is already present on the pane. Frost crystals are often referred to as hoar frost. When frost is severe and lasts for some time, it is sometimes referred to as black frost.

*Rime* is the name given to frost which occurs more commonly on mountain tops and on aircraft.

## Hail

The formation of hailstones takes place in stages. First, a little water condenses on a tiny solid particle in the air. This, upon meeting other particles, grows in size. The resulting droplet falls towards the ground. Second, a strong ascending current of air carries the raindrop aloft. If the level to which the droplet is carried by the updraught has a sufficiently low temperature, the droplet will freeze and fall as a hailstone. It is during summer thunderstorms when there are violent updraughts that this usually happens. In violent thunderstorms, the falling hailstones may be carried up again to form a further layer of ice. This may happen repeatedly, and the hailstone increases in size with each successive layer of ice formed.

## Water and Ice

When water is cooled, it contracts like any other liquid, but on reaching 4°C, it begins to expand. At 0°C, when it becomes solid, it expands about one tenth of its volume. This is why ice floats on water, and why there are eight-ninths of an iceberg below the surface and one ninth above it. This also explains why water at a lower level ceases to be cooled by means of convection currents once the temperature has dropped below 4°C. Water at the bottom of a pond, therefore, will remain at temperatures above freezing point, while that at the surface may reach temperatures many degrees below it. Life at the bottom of ponds and lakes can remain dormant but unfrozen. However, in a small outdoor container such as a jam jar, which is exposed to lowered temperatures on all sides, almost the whole mass of water may freeze owing to conduction.

Expanding ice is responsible for the burst pipe, the cracked jam jar, the breaking up of soil, and even cracks in rock and concrete.

## Melting

When any solid changes to the liquid form, it is said to melt. Melting is brought about by a rise in temperature and is not the same as dissolving. Melted ice and snow mostly finds its way back to the sea. Some of it sinks into the ground and drains into streams and rivers. A little escapes in the form of water vapour, to mix with the other gases in the air. Sometimes melting is followed by freezing. Frozen snow is snow whose surface has melted and then frozen into a crust of ice.

Where the climate is too cold for all the snow to melt in summer, the resulting accumulation may congeal and form streams of ice known as *glaciers*. Owing to gravity, these glaciers are forced down valleys into warmer zones, where sections melt off, or are pushed into the sea to form icebergs. The quantity of ice and snow in mountain and polar regions is such that, if it should all melt, the sea level would probably rise some 30 to 60 metres.

## Sleet

Sleet can be formed in two ways:
a snow falling from a high altitude may partly melt on the way down
b rain falling from a high altitude may partly freeze on the way down.

In Canada and America the term sleet is sometimes applied to tiny grains of ice or ice pellets which would, in Britain, be referred to as hail.

## CODE

1 Observe that frost found on a window pane in the morning is on the inside of the pane.
2 Observe the frost inside a refrigerator, or on the inside of a deep-freeze, *e.g.* in a shop where frozen foods are sold.
3 Observe that a lump of ice or a snowball floats on water.
4 If possible, observe a snowflake through a magnifying glass. It is better seen against a dark background, *e.g.* black card or cloth.
5 Demonstrate the formation of frost on a cold surface as a result of the freezing of water vapour in the air.
   a Pack a tall tin with alternate layers of crushed ice and salt. Use about twice as much ice as salt. Pack as tightly as possible.
   b Observe after a while the formation of frost on the outside of the tin.
6 Continue weather record as suggested in CODE 10, Lesson 11.

## Written Work

1 When a liquid freezes, it becomes a solid.
2 Ice is frozen water. Hailstones are frozen raindrops.
3 Snow is formed when water vapour freezes in the air.
4 Frost is formed when water vapour freezes on the land.
5 When a solid melts, it becomes a liquid.
6 Sleet is melting snow or freezing rain.

149

# 17

# A SOLID WE NEED

## COMMON SALT

### Demonstration Material

1   Common salt, *i.e.* cheap cooking salt
2   Two plant pots or the top halves of two plastic detergent bottles with holes plugged with cotton wool
3   Sand or aquarium gravel
4   Two jam jars or similar containers for collecting water
5   Saucers or tin lids
6   Water

### Sample Link Questions

1   What causes water vapour to change into a cloud of tiny drops of water? (*Cooling*)
2   What forces the clouds of drops of water to move over land? (*Moving air – Wind*)
3   What forces snow and rain to fall? (*Gravity – The pull of the earth*)
4   What does water flow through on its way to streams and rivers? (*The ground*)
5   When a solid melts, what does it become? (*A liquid*)
6   When a solid dissolves, what does it do? (*Mixes with something else*)
7   What is the common liquid in which many solids, liquids and gases dissolve? (*Water*)
8   What does gravity force stream and rivers to flow towards? (*The sea*)
9   What is the main difference between sea water and fresh water? (*Sea water has more salt in it*)
10  What do sun and moving air help water to become? (*Water vapour*)

### Relevant Information

The main points of this lesson are:

1   common salt is a solid we need;

150

2   it is carried in solution by water making its way through the land down to the seas.
3   We obtain it:
    a   by allowing sea water to evaporate
    b   by mining salt beds deposited by seas which dried up ages ago.

The chemist gives the term *salt* to a compound formed when the hydrogen of an acid has been replaced by a metal. Copper sulphate, for example, is a salt obtained from copper and sulphuric acid. The salt referred to in this lesson is, of course, common salt.

Common salt is a chemical obtained from sodium (a metal) and hydrochloric acid. It is a compound of the two elements sodium and chlorine and is known to the chemist as *sodium chloride*. Cooking salt which we buy in the shops is more or less pure common salt. Table salt is common salt mixed with other chemicals to make it pour more easily.

Salt is present in rock formations of all ages. When mixed with water, it dissolves readily. Along with other dissolved substances it is carried to the seas by streams and rivers. Thus the seas are gradually being made more salty by an agent other than the fairy-tale salt machine. In addition to this, the salt which is carried to the sea remains in solution. When water from the surface of the sea evaporates and becomes water vapour, the salt is left behind. The water vapour condenses into clouds. Rain falls from the clouds and sinks into the land, dissolving more salt on the way. This water makes its way to streams and rivers, and so back to the sea. The water cycle continues, but it carries salt on only part of its journey – from land to sea. There are minute particles of salt in the atmosphere, especially above the sea, but these do not result in any effective movement back to the land. Even so, the increase of saltiness in the sea is hardly noticeable. The salt which is present in stream and river water is not normally detectable by taste, and it has been estimated that it has taken something like 300 to 400 million years to make the present seas as salty as they are. It is the presence of salt in sea water which is responsible for its apparent colour in sunlight.

The concentration of salt in the Dead Sea is such that bathers cannot sink in it. The River Jordan adds something like 850,000 tonnes of salt each year, and the heat of the sun causes considerable evaporation.

Rock salt occurs in large deposits or 'beds' in various parts of the

151

world. These beds mark the position of seas which existed millions of years ago. The seas dried up, and the salt remained as a deposit to be covered eventually by layers of mud or sand. It is these underground deposits which are at present mined for a considerable amount of the salt used for commercial and domestic purposes today.

Salt is an essential part of the blood, and it is maintained in the blood in practically the same proportions as is the salt in the sea. The salt in the blood assists in dissolving protein, most of which will not dissolve in pure water. The body of the average human adult contains about 56 grams of salt.

Salt is lost from the body through the excretion of sweat and tears, and if neither of these is excessive, the loss of salt is small. Where there is excess sweating, caused for example by climate or certain kinds of employment, then it is necessary to add salt to the diet for health reasons. People who do heavy work in hot surroundings are apt to lose a considerable amount of perspiration. As this is virtually a pure solution of salt, the concentration of salt in the blood is liable to drop. If the salt content of the blood drops below certain levels, the result may be cramp in various muscles. So far as the requirements of the body are concerned, Europeans do not normally need additions of salt to their diet. Much depends, of course, upon the type of food consumed. When meat is roasted, for example, it retains its salt content, but when it is boiled it tends to lose it. Plant foods, too, call for a supplement of salt.

Human beings have, however, used salt for seasoning for as long as their history records. As it is scarce in some localities, it has long been used in commercial trading. It still constitutes a large part of the material carried by caravan traders across the Sahara desert. Roman soldiers were paid an extra allowance with which to buy salt. This was known as salt money (*salarium*), and from this word we get our world *salary*.

The two main sources of salt are:

1  Beds of rock salt, from which salt is mined or quarried by the usual methods. Rock salt is sometimes of a high degree of purity. Cheshire salt is, for example, 95% pure. Salt is also obtained from underground beds by forcing water in from above to dissolve the salt, pumping up the resulting solution, and then evaporating.
2  Sea water itself – brine. In every 45 kilograms of sea water,

there are about 1.6 kilograms of solid material in solution, and of this quantity of solid, some 1.2 kilograms is common salt. By allowing large areas of sea water to evaporate, this salt is obtainable. The usual method is to allow sea water to flow into specially constructed ponds known as pans, and after the evaporation of the water, to rake the salt into heaps until dry. This system is of course best in maritime countries with a dry climate and relatively long summers. It is not employed in Great Britain.

## Some Uses of Salt

The best known use of salt is of course as a seasoning. However, it is necessary to many industries, such as the soap, glass, and pickle industries, and is useful in packing meat and in curing fish and hides. It has long been established as preservative. It was used, for example, by seafarers for pickling their pork and by ancient Egyptians for embalming their dead.

In some parts of the world – for example, Central Africa – the scarcity of salt makes it a valuable luxury. Cakes of it have been used as currency in Tibet and other parts of the world.

Living things cannot stand an excess of salt. That is why it is useful in treating infections caused by bacteria or fungi. Fungus disease on goldfish, for example, may often be cured by adding salt to the water – sufficient to kill the fungus, but not enough to kill the fish.

## Customs, Sayings and Superstitions

The value of salt to life has brought it into sufficient prominence for it to be connected with many customs, sayings and superstitions. As long ago as Norman times, the lord of the manor sat at a raised table with a large salt container on it. Lesser mortals sat lower down the hall – thus the custom of 'sitting below the salt'. During the sixteenth and seventeenth centuries, many magnificent salt-cellars were made, and to 'sit above the salt' at table was a mark of class distinction. It is a custom in the East for the taking of salt at a meal to be regarded as binding a friendship between host and guest; in the story of Ali Baba for example, the robber chief, when dining in the guise of a merchant at Ali Baba's house, requested that his particular meal be unseasoned with salt, ostensibly for medical reasons, but in reality to avoid being bound by the custom. Salt has also been subject to tax, and the high tax imposed at the time was one of the causes of the French Revolution. Many sayings are associated with salt, *e.g.* 'to sit above the salt', 'worth his salt', and 'ye are the salt of the earth', etc.

It is not unnatural that salt should have become involved in superstition, and even today the spilling of salt at the table causes some people to throw a sprinkle over one shoulder in order to ward off evil spirits.

## CODE

1   Experiment to find how many teaspoonsful of salt can be dissolved in a jar of water. Count each one as it is put in, and then stir. Observe that although the dissolved salt can no longer be seen, it can be tasted.
2   Observe that perspiration and tears taste of salt.
3   Demonstrate how water filtering through the earth dissolves salt which is mixed with the rocks and soil.
    *a* Plug the holes of two plant pots with cotton wool.
    *b* Fill one with sand or aquarium gravel.
    *c* Fill the other with a mixture of salt and sand, or salt and aquarium gravel.
    *d* Pour water into both, and collect it in separate containers.
    *e* Compare the taste.

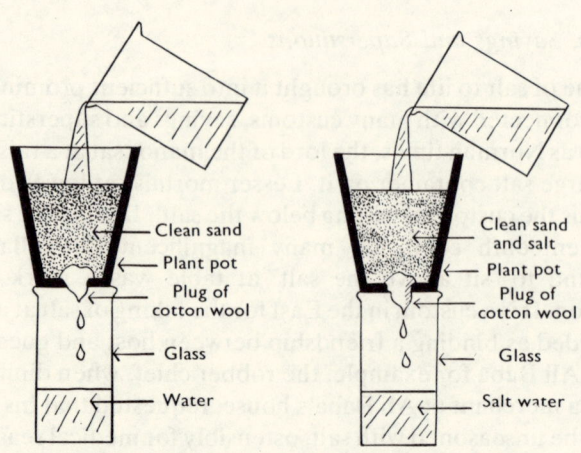

RAIN WATER DISSOLVES THE SALT IN THE GROUND

154

4  Experiment to find that when salt water dries up, the salt is left behind. Pour salt solution collected from Experiments 1 or 3 into separate tin lids. Allow the water to evaporate, and observe that salt is left.

## Written Work

1  Salt is <u>mixed</u> with rocks and soil.
2  Salt <u>dissolves</u> in water. Rivers <u>carry</u> it to the sea.
3  When sea water dries up, <u>salt</u> is left behind.
4  Table salt has other <u>solids</u> <u>mixed</u> with it.

# A GAS WE NEED

## OXYGEN GAS

## Demonstration Material

1  Either or both of the following:
   a small sections of candle,
   b metal caps from ink or other bottles, and liquid fuel, *e.g.* methylated spirit, paraffin, lighter fuel.
2  Several jars of different sizes, and tin lids on which to stand them.
3  Two jars of equal size, fitted with lids.
4  a Liquid fuel,
   b small tin lid,
   c sheet of exercise paper.

## Sample Link Questions

1  What are the four main needs of living things? (*Oxygen, food, to grow, to have young*)
2  Which two of these do living things need in order to grow bigger? (*Oxygen and food*)
3  What is air a mixture of? (*Gases*)
4  What do fish have to help them to breathe under water? (*Gills*)

5  What do we have to help us to take oxygen from the air? (*Lungs*)
6  Do air and water force themselves to move, or are they forced to move? (*They are forced to move*)
7  Which liquid do all living things need for food? (*Water*)
8  Name a solid that we need in our food. (*Salt*)
9  When we mix salt with water, what happens to some of the salt? (*It dissolves*)
10  What is the liquid in which many things dissolve? (*Water*)

## Relevant Information

The main points of a lesson on air in Pupils' Book 1 were:

1  air is something
2  we need it for breathing.

The main points of this lesson are:

1  oxygen is one of the chief gases in the air, and there is also some dissolved in water;
2  some animals have lungs to take oxygen from the air, and some animals have gills to take it from water;
3  oxygen is used when most things burn.

Oxygen is considered the commonest element on our planet. As it combines readily with most other elements, however, a considerable amount of it is always in combination with these other elements, forming various compounds. It forms, for example, one-third by volume of all the water in the world, and about eight-ninths by weight. It has been estimated that one half of the crust of the earth consists of oxygen. Oxygen which is not in combination with other substances is often referred to as free oxygen.

At normal temperatures, free oxygen exists in the form of a gas, and, as such, forms about one-fifth of the atmosphere (21% by volume, and 23.2% by weight). Free oxygen is also slightly soluble in water. At 15°C, for example, normal fresh water has an oxygen content of 1%, and sea water approximately 0.8%. This dissolved oxygen is used by fish and other living animals and plants which absorb their oxygen from water.

To living plants and animals, oxygen is perhaps the most important of the never-alive things. Energy is required by both animals and plants in order to utilise food for body building purposes, and living animals require additional energy to force themselves to move from one place to another. The energy for these and other cellular activities is obtained by both living animals and living plants through the oxidation of foodstuffs.

This absorption of oxygen from the air or water by living animals and plants is the beginning of the process known as *respiration*. It covers all the steps involved in the production of energy, from when oxygen, is first taken in, up to when it is given off combined with carbon, in the form of carbon dioxide.

## Breathing

The principle of diffusion of gases during respiration remains the same for all living things, but in the more highly evolved members of the plant and animal kingdoms various aids have been evolved to assist in the exchange. Although the term breathing is sometimes loosely applied to the absorption of oxygen by any living animal or plant, it is more generally restricted to those animals which assist the exchange of gases by an active movement of some part of the body. Where breathing takes place, a supply of air or water is continually forced into contact with a surface specially adapted for the rapid exchange of gases. Fish and certain other aquatic animals, for example, force a stream of water over special organs known as gills. Birds, mammals, reptiles and adult amphibians force air into special organs – the lungs.

Breathing is not associated solely with lungs and gills. An insect, for example, possesses neither of these parts, and yet may still assist respiration by mechanical means. In fact, respiration among insects is both efficient and unique.

An insect has a network of fine passages running all over its body. Through the walls of these passages (known as tracheal tubes) the exchange of gases takes place. The land-dwelling insects have tiny holes, known as *spiracles*, in the surface of the body, and through these spiracles air enters the internal passages. The larger insects in particular may be said to breathe, as they ventilate these passages by the mechanical expansion and contraction of the body wall.

Aquatic and parasitic insects have various adaptations for assisting respiration. Most aquatic insects obtain their oxygen from the atmosphere and have a variety of ways of bringing their spiracles into contact with the air above the surface of the water.

The larvae of some aquatic insects, however, obtain their oxygen from that which is dissolved in water. In such cases the tracheal tubes are closed, and oxygen diffuses into an extensive meshwork of branch tubes situated very close to the surface of the body. Some of these aquatic larvae, *e.g.* those of the dragonflies and mayflies, have special outgrowths containing these tubes. These outgrowths from the body are known as *tracheal gills*. In general, when aquatic insects reach the adult stage, they obtain their oxygen from the air, although there is one genus of bugs (*Aphelochirus*) which, as adults, absorb their oxygen from that dissolved in water.

Any living thing which is unable to obtain enough oxygen to meet its energy requirements, dies. Drowning and suffocation are examples of what happens when there is an insufficiency of oxygen. If, in a crowded room, the ventilation is not sufficient to allow an adequate supply of fresh air, we say that the air is stuffy, meaning that the oxygen is being used up more quickly than it is being replaced. The famous Black Hole of Calcutta was a classic example of death resulting from a lack of fresh air.

As the atmosphere is less dense at high altitudes, the amount of available oxygen is less. This is why equipment for supplying oxygen is worn at high altitudes. Such equipment was used, for example, during the first conquest of Everest in 1953. Equipment for containing and supplying oxygen is of course a major consideration during journeys into space.

It would seem that rest and sleep are associated with oxygen requirements. While we are awake, we tend to use up more oxygen than when we are asleep.

### Burning

Oxygen combines readily with most of the other elements. When the temperature is raised, it combines more rapidly. The process of combining with other elements is known as *oxidation*. When light as well as heat results from oxidation, it is termed *combustion* or burning. A flame is generally understood to consist of gases which are burning. A flame may not consist entirely of burning gas however. For example, a candle flame is really a shell of flame which contains glowing particles of solid carbon.

When a substance is burning due to oxygen combining with it, the obvious way to stop the burning is to deprive it of further oxygen. Covering such a burning substance with water tends to prevent further combustion, because the water cuts off the supply of free oxygen which is in the air. The slight amount of free oxygen that is dissolved in the water is insufficient for burning; and the oxygen that is chemically a part of the water is, of course, not in a free state.

Liquids which float on water cannot be so easily stopped from burning. Obviously they will tend to float on any water which is directed at them and so will continue to have access to the free oxygen in the air. Water is therefore of little use in putting out fires consisting of burning fats or oils, and other means have to be found of smothering the flames.

Many substances, such as sand and chalk, will not burn because they are already oxidised. The rapid combination of some substances with oxygen may result in an explosion.

As burning is any chemical reaction during which heat and light are evolved, it is possible to have burning without oxygen being present. For example, heat and light may result when hydrogen and chlorine combine to produce hydrochloric acid. Such a reaction can take place without oxygen being present.

## Where does all the oxygen come from?

Although plants absorb oxygen during respiration, just as animals do, certain of them are responsible for releasing free oxygen to the air or water in which they live. These are the 'green' plants, *i.e.* the ones which contain chlorophyll. Plants which contain chlorophyll are able to feed by a process known as *photosynthesis*. During the process of photosynthesis, a plant utilises the carbon from carbon dioxide gas. As carbon dioxide is a compound of carbon and oxygen, the oxygen is set free during the process. This sometimes gives rise to the false statement that plants 'breathe out' oxygen. What actually happens, is that in sufficient light, the process of photosynthesis masks that of respiration, *i.e.* during its feeding process, the plant is liberating more oxygen that it is using for respiration. It is this excess which maintains a constant supply of free oxygen for the remainder of the plant kingdom, and the whole of the animal kingdom.

*Green* plants include all the herbs, trees and shrubs, and such simple plants as the alga class, and the class of mosses and liverworts.

Although underwater plants containing chlorophyll supply free oxygen to be dissolved in the water, oxygen also diffuses in through the surface. This diffusion through the surface is not rapid however, and it follows that for a given volume of water, a large surface area permits a more rapid exchange of gases than a small surface area. This is why a goldfish globe is a most unsuitable home for fish, and why fish may be seen 'gasping' at the surface when the container which they occupy is overcrowded.

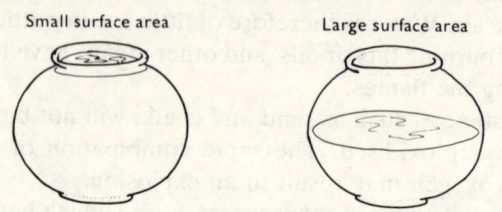

Small surface area        Large surface area

A GOLDFISH BOWL IS AN UNSUITABLE HOME FOR A GOLDFISH

During breathing, a fish absorbs oxygen which is dissolved in the water. In consequence of this, more oxygen diffuses in through the surface of the water, from the air. The process is not rapid, however, so that the greater the area of water surface in contact with the air the better. To permit maximum absorption, the globe would need to be about half empty, thus restricting the room for swimming.

The non-human animals illustrated in Lesson 18 of the Pupils' Book are the herring and the larva (caterpillar) of the privet hawk moth.

## CODE

1 Observe the need to open windows in a stuffy classroom in order to allow a more plentiful supply of oxygen to replace that which is being used up by the class.

2 Observe any examples of fish in a crowded container 'gasping' at the surface for the oxygen which diffuses there into the water – for example, over-crowded goldfish in an aquarium in a pet shop, or fish kept in a globe or other container with a narrow neck.

3 Observe the ventilation holes in a grate to allow the passage of fresh air containing oxygen to the fire. Observe also that when the holes are open, the fire burns more quickly, and that when the holes are closed, the fire burns more slowly.

4 Experiment with a control to show that burning cannot continue after the available oxygen has been used up.

   a Light two candles, and invert a glass jar over one. Observe that the enclosed candle burns for only a short time – until it has used up the available oxygen.

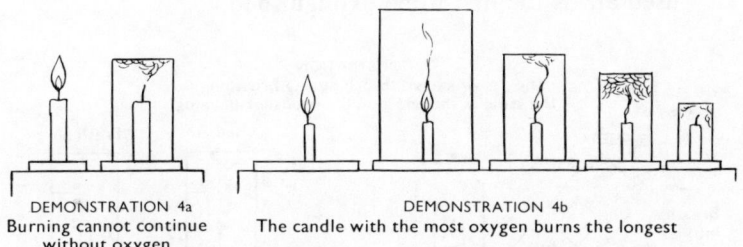

DEMONSTRATION 4a
Burning cannot continue
without oxygen

DEMONSTRATION 4b
The candle with the most oxygen burns the longest

   b Take several pieces of candle of the same length, and the same number of glass jars, minus one. The jars must be of noticeably different capacities.

   Stand the candles in line and light them.

   With helpers, invert jars over all the candles but one, *simultaneously*. It helps if the jars are in order of size along the line, with the largest next to the uncovered candle.

   Observe that the order in which the flames are extinguished coincides with the sizes of the jars, *i.e.* depends on the amount of oxygen available.

   c Quickly invert over another lighted candle one of the jars under which a candle has burned. Observe that this candle flame is extinguished immediately (owing to lack of oxygen).

*Notes*

   i These experiments may be conducted with similar-sized ink-bottle caps containing a liquid fuel, such as methylated spirit or lighter fuel, instead of candles.

   ii Small tin lids provide a smooth surface on which to stand candles and jars.

5 Experiment with a control to show that during breathing, we absorb oxygen from the air.

   a Obtain two glass jars of the same size, and fitted with lids, *e.g.* Kilner jars.

*b* Have a child breathe out several times into one of the jars, replacing the lid quickly after each 'breath'.

*c* Have the lids removed and the two jars inverted *simultaneously* over lighted candles (or small amounts of burning liquid fuel).

*d* Observe that the candle (or liquid fuel) in the jar containing 'used air' is the first to be extinguished.

DEMONSTRATION 5
The gas which we absorb during breathing is
the same as the one a candle needs for burning

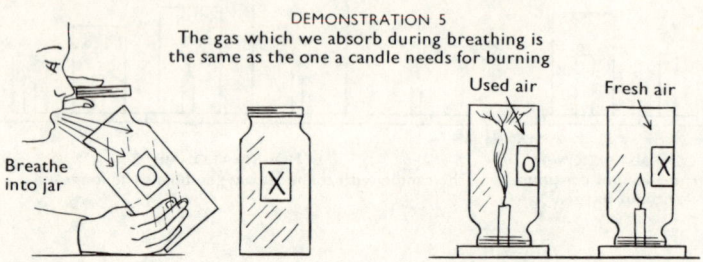

Breathe into jar    Used air    Fresh air

6 Demonstrate that burning ceases when the oxygen supply is cut off by 'smothering'.

   *a* Pour a small quantity of liquid fuel, *e.g.* methylated spirit, into a small tin lid and light it.

   *b* Hold a sheet of exercise paper between both hands; bring the paper down smartly, so as to completely cover the rim of the tin lid. Observe that the flame is completely extinguished.

**Written Work**

1 Some oxygen <u>gas</u> is <u>dissolved</u> in water.
2 Gills take oxygen from the <u>water</u>.
3 Lungs take oxygen from the <u>air</u>.
4 During breathing, air or <u>water</u> is <u>forced</u> to move.
5 Most things use <u>oxygen</u> when they <u>burn</u>.

# FORCING THINGS TO MOVE

## MAGNETS PULL

**Demonstration Material**

1 Magnets of various sizes and shapes, *e.g.* straight bar magnets, horse-shoe magnets, U-shaped magnets, rod magnets, etc.

2 Assorted oddments which can be forced to move by an ordinary magnet, *e.g.* hair clips, paper clips, tin lids, drawing pins (steel plated with brass), steel nails, screws, needles and *panel pins*. (Iron filings tend to be messy.)

3 Assorted oddments which cannot be forced to move by an ordinary magnet, *e.g.* sand, chalk, cloth, pebbles, wood, cardboard, cork, polystyrene, plastic items, paper, glass marble, copper, aluminium, zinc, brass, etc.

4 Any or all of the following:
  *a* small tin floating in a bowl of water
  *b* steel ball bearing with a thin covering of modelling clay
  *c* strong magnet, cellotape, sewing needles and thread
  *d* materials for making a magnetic force measurer (See CODE)
  *e* unmagnetised nail or long steel needle.

**Sample Link Questions**

1 Which kind of living things can force themselves to move from place to place? (*Living animals*)

2 What forces the clouds to move? (*Wind – moving air*)

3 What forces water to fall from the clouds? (*Gravity*)

4 Is gravity a pull or a push? (*A pull*)

5 What makes carrying heavy things harder than sliding or rolling them? (*Gravity*)

6 Is the earth the only thing with gravity? (*No, all things have gravity*)

**Relevant Information**

The main points of this lesson are:

1 magnets have a pull which can force some things to move. We call this pull MAGNETISM;

2   the magnets which we normally use are bars – either straight or
    bent – but they can be of any shape;
3   the pull is greatest at the magnet's poles – which are near its
    ends;
4   iron and steel are the materials most affected by an ordinary
    magnet but, so far as is known, any material would react to a
    force of magnetism which was sufficiently strong.

North-seeking poles and south-seeking poles are the subject of a
separate lesson in Book 3, as are the uses of magnets and direction
cards in magnetic compasses.

So far as is known, all substances will react to a magnetic field,
provided that field of force is sufficiently strong. Some substances,
*e.g.* glass, bismuth and antimony, would be repelled from such a
field of force, and are known as *diamagnetic*. Others would be
drawn into such a field of force, and are known as *paramagnetic*.

The diamagnetic substances and most paramagnetic substances
will react only to a magnetic pull which is tremendously strong, such
as that of an extremely powerful electromagnet. Some paramagnetic substances however, are distinguished by the relatively considerable magnitude of their reaction to a magnetic field. Such exceptionally magnetic substances are known as *ferromagnetic*.

Prominent amongst these ferromagnetic substances are iron,
cobalt, nickel, and their alloys and compounds. Iron and its compound, steel, are perhaps the most commonly used and best known
of these. Cobalt and nickel are less magnetic than iron. When a
quantity of one of these ferromagnetic substances exhibits a magnetic force strong enough for its influence on other substances to be
observed, it is termed a *magnet*.

In a magnet the force of attraction is concentrated at two points
known as the *poles*. A magnet may be of any shape. Specially
manufactured magnets are usually in the form of a bar – either
straight, or bent into the shape of a U or a horseshoe. Such manufactured magnets are sometimes referred to as permanent magnets,
although, strictly speaking, they cannot be considered permanent.
In bar magnets – straight or bent – the poles are situated near the
ends of the bar.

Natural magnets in the form of a magnetic iron ore are found in
various parts of the world. This ore is a black oxide of iron –
*magnetite*. Magnetite is also known as lodestone (or loadstone), and
its magnetic properties are believed to have been known to the

Chinese as long ago as 2400 B.C. It is found in various parts of the world, and in particular Sweden, where it occurs in large quantities. Generally speaking, however, lodestone has less power than a good bar magnet.

When children are experimenting to find out which things react to an ordinary magnet, they should obviously obtain positive results from those made out of iron and steel. The iron and steel themselves are not always visible, however, as some manufactured articles may be plated with another metal to protect them against rusting. Steel drawing pins, for example, may be plated with brass, and 'tins' are often made of iron with only a thin coating of tin. Consequently, when other metals appear to react to an ordinary magnet, it is due to the concealed ferromagnetic substances, *e.g.* iron or steel, which they contain.

On the other hand, it will be found that some articles will react visibly to an ordinary magnet, while others which are similar will not. For example, some screws will show an obvious reaction because they are made of steel, while others will show no apparent reaction at all, because they are made of brass.

Often only the north-seeking pole is marked, but this does not mean that the magnet has only one pole. It is not possible to have a magnet with only one pole. Breaking a magnet into two, for example, will merely result in two smaller magnets, each with its own north-seeking and south-seeking poles.

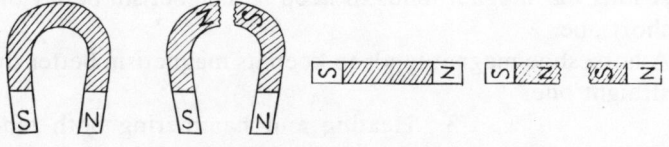

No magnet has only one pole

## Making Magnets

A method of magnetising, say, a strip of iron or steel is to stroke it a number of times in one direction only, with one pole of a magnet. About twenty strokes may suffice.

Soft iron is more easily magnetised than steel, but tends to lose its magnetism quickly when the magnetising force is removed. Steel retains its magnetism more tenaciously, when the magnetising force is removed.

Any number of magnets may be made from one magnet, without any loss of magnetism in the original magnet.

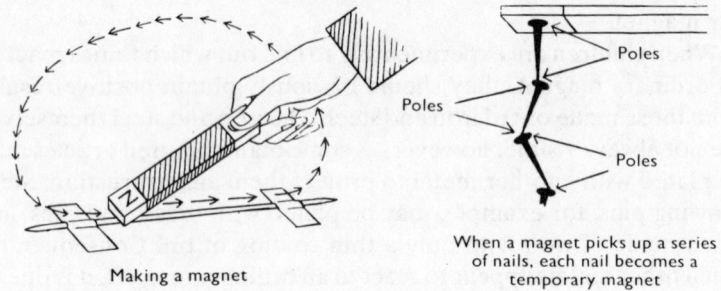

Poles

Poles

Poles

Making a magnet

When a magnet picks up a series of nails, each nail becomes a temporary magnet

When a magnet picks up a 'chain' of, say, tacks, each tack becomes magnetised by induction and becomes temporarily a tiny magnet. Each tack will have its own two poles, and it may be found afterwards that one or more of the tacks will remain slightly magnetic.

The strongest magnets are now made by using an electric current. There is a limit, however, to the force of magnetism which a magnet can acquire. This is known as the saturation point, and is relative to the size of the magnet.

## Care of Magnets

1 A long bar magnet tends to keep its magnetism better than a short one.
2 A horseshoe magnet tends to keep its magnetism better than a straight one.

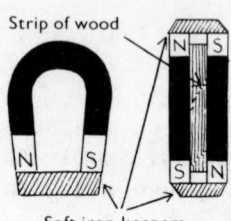

Strip of wood

Soft iron keepers

3 Heating and hammering both tend to demagnetise, unless the magnet lies along the lines of the earth's magnetic force, in which case hammering will tend to magnetise.
4 Soft iron loses its magnetism easily, steel less easily, and cobalt steel less easily still.

Complex alloys capable of withstanding the general effects of long usage are used in the manufacture of good quality magnets. Such magnets are the best for general school use.

166

5   Dropping magnets or packing them with their like poles together tends to weaken their strength. They are more stable if stored with the keepers supplied by the makers.

## CODE

1   Observe that magnets can be of different sizes and shapes – as well as horse-shoe shaped, *e.g.* straight bar, ring-shaped, U-shaped, bridge-shaped, rod-shaped, and otherwise shaped.

2   Experiment to find that an ordinary magnet can force some things to move but not others:
   *a* apply a magnet to a variety of materials, *e.g.* chalk, pieces of cloth, plastic materials, rock, wood, cardboard, glass and various metals
   *b* separate into two sets labelled MOVED and NOT MOVED
   *c* observe that the things which can be forced to move by an ordinary magnet are made of iron or steel, or have iron or steel parts; observe also that some of the metals are in the NOT MOVED set, *e.g.* brass, copper, zinc, aluminium, etc.

3   Experiment to find that the pull is strongest at the ends (about the poles) and weakest in between the poles:
   *a* dangle a sewing needle on a thread above the centre of a magnet, and slowly lower it; observe to which part of the magnet it is pulled
   *b* find out which part of the magnet collects the most tacks or panel pins – or from which parts most can be hung (steel panel pins and tacks are far less messy than iron filings).

4   Experiment to find that both ends attract the same things:
   *a* with materials from Experiment 2
   *b* float a small tin or tin lid in a bowl of water, and force it to move without touching it. Use each end of the magnet in turn.

5   Experiment to find that the pull of a magnet can pass through things:
   *a* place a steel paper clip or some other small object made of iron or steel on top of a thin surface, *e.g.* paper, card, plastic lid, etc. Move a magnet about underneath the surface.
   *b* cover a steel ball-bearing with a thin layer of modelling clay. See if a magnet can force this apparent ball of clay to roll.

6   Experiment to find if magnetism can overcome gravity:
   *a* force a small iron or steel object to move up a sloping surface, if possible without the magnet touching it.

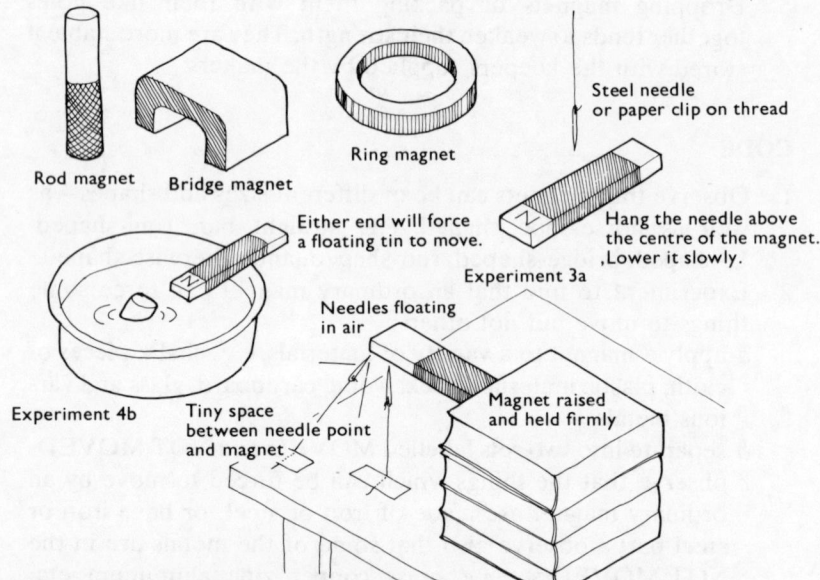

Rod magnet  Bridge magnet

Ring magnet

Steel needle or paper clip on thread

Either end will force a floating tin to move.

Hang the needle above the centre of the magnet. Lower it slowly.

Experiment 3a

Experiment 4b

Needles floating in air

Tiny space between needle point and magnet

Magnet raised and held firmly

Start with the needle point touching the magnet. Pull the thread slightly so that there is a tiny space between needle point and magnet. Fasten end of thread with tape. Paper clips may be used instead.

Experiment 6b MAGNETISM overcoming GRAVITY (or the Indian Cotton Trick)

    *b* if a sufficiently strong magnet is available, make a needle or paper clip float in the air on the end of a length of thread.

7    Experiment to find which of two magnets has the stronger pull. Place a tin lid between them, and pull slowly apart. See which magnet keeps the tin lid.

8    Demonstrate how to make a simple magnetic force measurer for measuring the pull of different magnets:

    *a* to the side of a tobacco or similar tin, attach one end of a 50 to 60 centimetre length of flat elastic

    *b* put sufficient sand in the tin to cause the elastic to begin to stretch. Secure the tin so that the sand cannot escape

    *c* fix a pointer to the tin, and hang down the wall alongside a home-made scale, so that the pointer is level with the zero mark on the scale

    *d* to use the magnetic force measurer, place a magnet against the lowest part of the tin, and pull it downwards, until the elastic forces it to spring back. Observe how far down the scale the pointer is pulled.

*Note*

Some of the ways in which this magnetic force measurer may be used are as follows:

*a* to find whose magnet has the strongest pull

*b* to find which of the two poles of an individual magnet has the stronger pull

*c* to show the advantage of a horse-shoe magnet, *i.e.* that both poles pulling the same object have a stronger pull than one pole on its own.

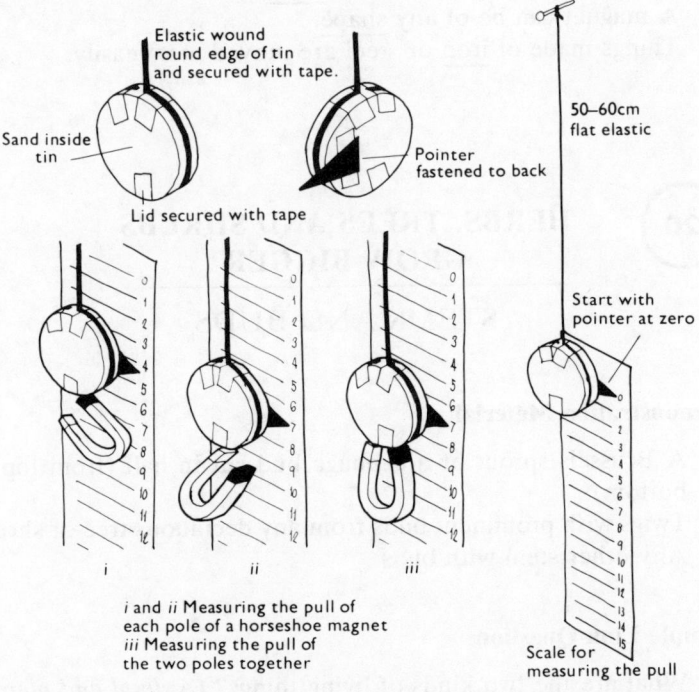

Elastic wound round edge of tin and secured with tape.

Sand inside tin

Lid secured with tape

Pointer fastened to back

50–60cm flat elastic

Start with pointer at zero

*i* and *ii* Measuring the pull of each pole of a horseshoe magnet
*iii* Measuring the pull of the two poles together

Scale for measuring the pull

9   Demonstrate that magnets can be made from materials which can be forced to move by a magnet:

   *a* take an unmagnetised nail or steel needle

   *b* show that this will not attract panel pins

   *c* demonstrate that it itself is attracted by a magnet

   *d* magnetise, as shown in illustrations on page 166

*Note*

Do not rub the needle with the magnet as is sometimes suggested.

Stroke the needle in one direction only, lifting the end of the magnet well clear of the needle for the return journey.

    *e* show that the steel needle will now attract steel panel pins to both ends, but not the section joining them.

## Written Work

1  A magnet's pull is called <u>magnetism.</u>
2  A magnet's pull is greatest at its <u>poles.</u>
3  A magnet's poles are near its <u>ends.</u>
4  A magnet can be of any <u>shape.</u>
5  Things made of <u>iron</u> or <u>steel</u> are moved most easily.

# 20  HERBS, TREES AND SHRUBS GROW BIGGER

## STEMS AND BUDS

### Demonstration Material

1  A Brussels sprout or a cabbage bud cut in half (from top to bottom).
2  Twigs with prominent buds from any deciduous tree or shrub.
3  Any other stem with buds.

### Sample Link Questions

1  What are the two kinds of living things? (*Animal and plant*)
2  What do we call plants which have no true roots, stems or leaves? (*Simple plants*)
3  What are the three kinds of plants with roots, stems and leaves? (*Herbs, trees, shrubs*)
4  What is the difference between the stems of herbs, and the stems of trees and shrubs? (*Trees and shrubs have woody stems*)
5  What is the difference between a tree and a shrub? (*A tree has one main stem growing from the roots. A shrub has more than one*)

6 What do we call trees and shrubs which have green leaves in winter? (*Evergreens*)
7 What name do we give to trees and shrubs which have no green leaves in winter? (*Deciduous*)
8 What do we call leaves which grow from the stem in twos? (*Opposite*)
9 What do we call leaves which grow from the stem in ones? (*Alternate*)

## Relevant Information

Buds are specialised growing points. The main purpose of this lesson is to establish that
1 buds grow from the sides and the ends of stems
2 the three different structures which develop from buds are:
   *a* sections of stem bearing leaves
   *b* flowers
   *c* sections of stem bearing both leaves and flowers.

The different ways in which buds are arranged on stems is the subject of a separate lesson in Book 3.

According to the position in which they grow, buds are usually named as follows.

1 *Terminal* – at the tip of the stem.
2 *Axillary* – in the axil of a leaf, which is the upper angle between the leaf and the stem.
3 *Adventitious* – at any point on the stem, and occasionally from a root or a leaf. When a tree has been cut down, for example, new branches may develop from adventitious buds on the stump.

Most adventitious buds are found on the sides of stems, as are the axillary buds. Buds which occur on the sides of a stem (side buds), are also referred to as *lateral* buds.

Terminal, axillary and adventitious buds are fundamentally similar in structure. On most species of plants they are distinguished solely by their position.

Sometimes the terminal bud continues the growth of the plant, and sometimes the side buds continue the growth. Horticulturists dictate which shall do so when they practise pruning.

Sometimes the terminal bud is larger than the side buds, but not always so. It should not be assumed, however, that the bud which is present at the end of the stem is the terminal bud. On some stems the growth of the stem may be sympodial, that is to say that growth may be continued by the successive development of side buds from just below the apex of the stem. In such cases the bud observed at the end of the stem may be a side bud, and not a terminal bud. This may often be found on stems where the normal arrangement of the side buds is an alternate one.

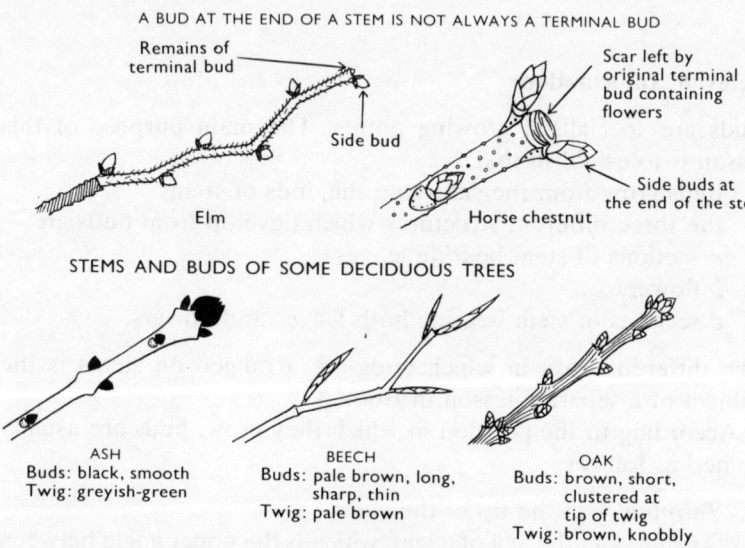

A BUD AT THE END OF A STEM IS NOT ALWAYS A TERMINAL BUD

Remains of terminal bud

Side bud

Elm

Scar left by original terminal bud containing flowers

Side buds at the end of the stem

Horse chestnut

STEMS AND BUDS OF SOME DECIDUOUS TREES

ASH
Buds: black, smooth
Twig: greyish-green

BEECH
Buds: pale brown, long, sharp, thin
Twig: pale brown

OAK
Buds: brown, short, clustered at tip of twig
Twig: brown, knobbly

The three main structures of buds are:

1 *Leaf Bud*. (This may be more accurately termed a stem or branch stem bud.) From it develop sections of stem, bearing new leaves.
2 *Flower Bud*. From this develops one or more flowers. On coniferous trees and shrubs it is a cone which develops. (A cone is the structure from which in time the flower evolved.) Cones and flowers are looked upon as being highly specialised twigs.
3 *Mixed Bud*. From this kind of bud develops a short length of stem bearing both leaves and flowers.

When a flower appears at the end of a stem, the growth of that particular stem is terminated.

Where there is continuous growth of the stem, as in those herbs, trees and shrubs which have no inactive period forced upon them, the buds are unprotected; but on trees and shrubs which experience a period of rest, as do those of temperate regions, the buds are protected by *bud scales*. Bud scales are modified leaves.

Deciduous trees and shrubs, with their leafless woody stems persistent above the ground during winter, are amongst the best plants to study for generalisations of bud arrangement and structure. On most trees and shrubs, buds form during the summer, so that the following year's leaves are ready in their buds by the early autumn. The waterproof bud scales protect them against winter weather and against loss of water. Some bud scales have hairy outgrowths to assist in this purpose, and others, such as the horse chestnut, have a coating of sticky gum.

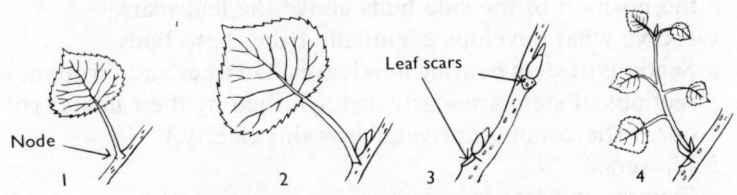

1 Leaf growing from node on branch stem in spring
2 Bud developing in axil of leaf during summer
3 Winter appearance, showing scar where leaf was attached
4 Branch stem growing from bud during the following
    year (in this case bearing leaves only)

Not all the buds on the stem of a tree or shrub should be expected to develop in the spring. Some remain dormant and inactive for many years, and some may never develop at all. Others may be killed by frost.

STEMS AND BUDS OF SOME EVERGREEN TREES

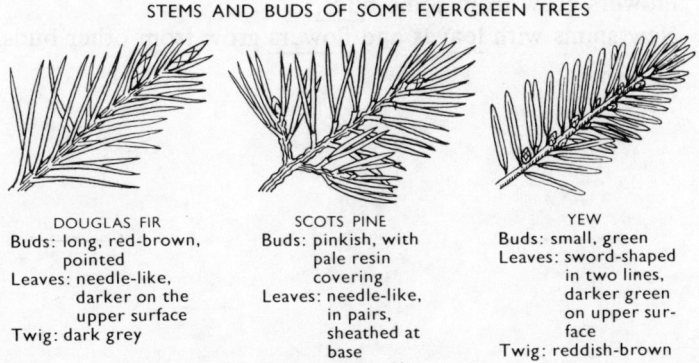

DOUGLAS FIR
Buds: long, red-brown,
    pointed
Leaves: needle-like,
    darker on the
    upper surface
Twig: dark grey

SCOTS PINE
Buds: pinkish, with
    pale resin
    covering
Leaves: needle-like,
    in pairs,
    sheathed at
    base
Twig: copper-red

YEW
Buds: small, green
Leaves: sword-shaped
    in two lines,
    darker green
    on upper sur-
    face
Twig: reddish-brown

173

**CODE**

1  Cut a sprout bud down the middle – or, better still, a cabbage –
   and observe:
   *a* compressed branch stem
   *b* leaves growing from that stem
   *c* further side buds growing in the axils between leaf and stem.
2  Collect the 'winter' twigs of any deciduous trees and shrubs.
   Crush or split the ends of these twigs before standing in water.
3  Observe on these twigs:
   *a* end bud and side buds
   *b* how the end (terminal) buds are larger than the side buds on
      certain twigs, *e.g.* horse chestnut and ash
   *c* the position of the side buds above the leaf scars.
4  Observe what develops eventually from these buds.
   *a* Sections of stem bearing new leaves. On trees and shrubs new
      sections of stem are easily distinguished by their lighter col-
      our. (The common privet shows this clearly.)
   *b* Flowers.
   *c* Flowers and leaves.
5  Observe how certain buds may not develop at all.
6  Observe the position of buds on the stems of herbs, and evergreen trees and shrubs, especially those found in leaf axils.


**Written Work**

1  Buds grow from the sides of a stem and at its end.
2  Most side buds grow where a leaf joins a stem.
3  New stems with new leaves grow from some buds.
4  Flowers grow from some buds.
5  New stems with leaves and flowers grow from other buds.

**LIVING THINGS HAVE YOUNG**

## THE ANIMALS

### Demonstration Material

Any or all of the following:
1 Animals which can reproduce from part of the parent, *e.g.*
   *a* starfish, sea anemone, freshwater hydra
   *b* skeleton remains – coral, sponge, sea-mat.
2 Egg or egg cases, *e.g.*
   *a* bird's egg
   *b* insect eggs on a plant part
   *c* snail eggs
   *d* egg case of dogfish
   *e* egg case of skate
   *f* egg cases of whelk.

### Sample Link Questions

1 What are the four main needs of living things? (*Oxygen, food, to grow, to have young*)
2 What are the two kinds of living things? (*Animal and plant*)
3 Are seeds and eggs alive or dead? (*Alive*)
4 What grows from an egg – a young animal or a young plant? (*A young animal*)
5 What are the stages in the life of an insect with a three-stage life? (*Egg, larva, adult*)
6 What are the stages in the life of an insect with a four-stage life? (*Egg, larva, pupa, adult*)
7 Which insect lays the eggs – the male adult or the female adult? (*The female adult*)

### Relevant Information

The fundamental aim of a species is the propagation of its own kind. If, in a particular species, this aim were to be abandoned or unfulfilled, that species would be on the way to extinction. The main purpose of this lesson is to show that there are three main ways in which animals reproduce themselves:

1   by the development of a new individual animal from a section of
    the parent
2   by the laying of eggs
3   by giving birth to one or more individual animals which are not
    contained in an egg.

### 1   *The Development of a New Individual from Some Section of the Parent*

In this method part of the parent animal develops as a new individual and usually becomes detached from the parent during
growth. In other instances the individuals may remain attached,
resulting in the formation of colonies; examples are sponges, corals
and sea-mat. Although this method of reproduction is not encountered quite so extensively in the animal kingdom as it is in the plant
kingdom, it occurs amongst the lower levels of the animals. For the
simplest animals it is the only method of reproduction possible.

### *a One-celled animals*

A one-celled animal simply divides, one individual becoming two.
The amoeba is the best known textbook example of this, but there
are many others. This is the method by which all forms of life grow
in size, the individual cells of which they are composed increasing

How the cells of living things increase

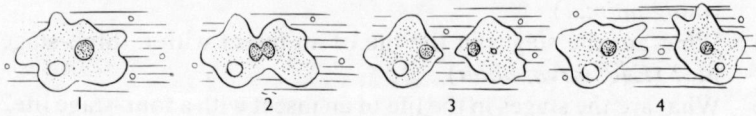

their number by dividing (multiplication by division). In one-celled
animals there would appear to be no death from old age. If death
does occur, it is because of lack of food, attack by enemies, or
because there is something wrong with the environment.

### *b Animals consisting of more than one cell*

Certain multi-celled animals are capable of this kind of reproduction. In some instances – for example, the sea anemone – the entire
body may split into two. In other cases the body may separate into

sections, each of which may grow into a new individual – for example, certain flatworms, annelid worms, sea cucumbers and starfish. Certain starfish break apart at fairly regular intervals of time. Even if a broken-off ray has no part of the central disc attached to it, it is possible for a whole new starfish to develop from it.

The common flatworms or planarian worms of ponds and streams are also remarkably capable of developing as new individuals from any fragment of the original animal. The planarian worm has eyes and a brain, yet any section cut from any other part of the body will be capable of developing as a new individual, complete with eyes and a brain.

How sections of a planarian may develop into new individuals

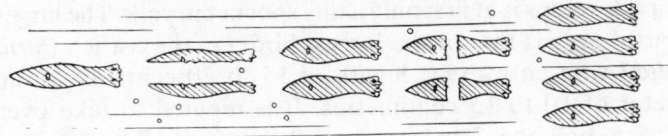

The freshwater hydra of ponds and streams is capable of developing new individuals from mere fragments taken from any part of the body. It is also an example of an animal from which a new individual may develop as an outgrowth from the parent animal – a process sometimes referred to as *budding*.

How sections of hydra may develop into new individuals

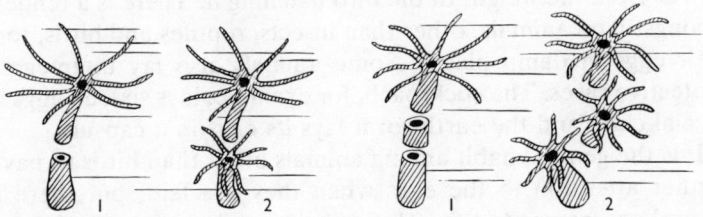

## 2   *The Laying of Eggs*

Reproduction by means of eggs is far superior in many ways to methods involving subdivision of the animal itself. It is much the commonest method in the animal kingdom. Multi-celled animals such as the starfish, the hydra, the sea-anemone and the planarian worm, which can reproduce by means of subdivision, may also reproduce by means of eggs.

177

The number of eggs laid varies from species to species. Generally speaking, the quantity varies according to the risks of destruction, the number being greatest where losses are liable to be greatest. Some animals lay only one egg a year; others lay millions. The common oyster lays some 60 000 000 in a year, and queen termites are reputed to lay over 4 000 a day for years.

The shape of eggs lies between the ovoid and the spherical but, especially among insects, there is an interesting variety of forms.

The size of an egg depends upon the amount of food it encloses. The eggs of some animals contain little food and are very tiny. The eggs of reptiles and birds contain large amounts of food, and are relatively large. Birds' eggs are generally the largest. In these, the white and most of the yolk is a food supply, and the germ from which the bird develops is at first only a tiny spot in the yolk. The largest of the birds' eggs is that of the largest bird, *i.e.* the ostrich (*Struthio camelus*) with an average length of 15 to 20 centimetres, and a diameter of 10 to 15 centimetres. It is reputed to take over 40 minutes to boil one. The largest of all eggs (including egg cases) in the animal kingdom, however, is that of the largest fish – the whale shark (*Rhiniodon typus*). On record is a whale shark egg case with the dimensions of $30 \times 14 \times 9$ centimetres.

The outer coverings of eggs vary according to the conditions to which they are exposed. Eggs which are laid in water, for example, do not need protection against drying up, while those laid on land do. The limestone shell of a bird's egg protects it from drying up, as well as from the weight of the bird hatching it. There is a tendency amongst land animals, other than insects, reptiles and birds, to lay their eggs in damp places. Some animals also lay their eggs in protective cases. The cockroach, for example, lays sixteen eggs in a special case, and the earthworm lays its eggs in a capsule.

It is the general habit among animals other than birds to pay no further attention to the eggs when they are laid; but there are several exceptions to this. The midwife toad carries its eggs on its back, and certain crustaceans – for example, the shrimp – carry their eggs about with them. Quite a number of animals, however – particularly insects – go to some trouble to lay their eggs close to, on, or actually inside an appropriate source of food. Some animals also lay their eggs where they are protected from enemies. Birds build nests for this purpose. Also, on the majority of birds' eggs are the various colourful tints which assist in camouflage.

The salmon and the eel are examples of animals that overcome considerable hazards in order to lay their eggs in suitable places. The salmon, which is a salt-water fish, spends some time in the mouth of a river adapting itself gradually to freshwater conditions, and then undertakes the epic journey up rivers and waterfalls to lay its eggs in some quiet stream. The eel follows the reverse procedure, making its way from fresh water to the sea – even crossing land *en route* – until eventually having acclimatised itself to salt water, it is able to return to the Sargasso Sea where the eggs are laid. Such peculiarities are of course exceptional.

The majority of eggs require fertilisation before development can take place. Hens' eggs which are sold for culinary purposes are (or should be) unfertilised eggs. There are some exceptional instances of animals whose young develop from eggs which have not been fertilised by a male cell. This is encountered among species of stick insects kept in Britain.

### 3 The Birth of Individual Animals which are not Contained in an Egg

This is an operation sometimes referred to as 'live-bearing', to distinguish it from egg-laying. The term is misleading, however, as fertilised eggs also have life in them when they are laid. Animals which are not laid as eggs have certain advantages, in that they have undergone the early stages of their development in relative safety, and their chances of survival are greater.

This method of reproduction has been most extensively developed among mammals. The class of mammals is divided into three sub-classes:

1   the primitive egg-laying mammals – the duck-billed platypus and the echidnas
2   the marsupials, or pouched mammals – the kangaroos, wallabies, wombats, opossums and koala bears
3   the placental mammals.

Like birds, mammals evolved from reptiles, but whereas reproduction continues to be by means of eggs in the bird class, it is confined in the mammal class to the most primitive species. Egg-laying mammals and the marsupials live mostly in Australia and New Zealand. Marsupial mammals are born relatively undeveloped, and they therefore need to be carried about by the parent for some time after birth. In most cases they are carried in a fold of the skin known as a pouch.

The majority of the mammals belong to the third sub-class – the placentals. In some cases the young are comparatively helpless at birth and unable to move very far. This applies to such mammals as kittens, puppies, hamsters, rabbits, and humans. The young of some other mammals, such as horses, sheep, cows and hares, on the other hand, are very active and able to move about quite freely almost as soon as they are born.

Among fish, insects, reptiles, and snails there are exceptional species which do not *lay* eggs, but retain them internally until after the young have hatched out. The number of active young which such a parent can produce in this way is obviously less than the number of eggs which could be produced. However, as the young are active and able to move from place to place, their chances of survival are greater, and therefore the actual wastage is less than with eggs.

Three of the most popular fish kept in tropical aquaria in homes and schools are the common guppy, the swordtail and the platyfish. All three give birth to free-swimming young and in consequence are very easy to breed. The common ked or sheep tick is an insect which retains the larvae almost until the stage of pupation, thus earning for itself the name of *pupa bearer*

### Care of Young

It is only amongst mammals and birds that taking care of the growing young is the general practice. Mammals are the only animals whose young must be fed on milk, and it is necessary for the parent to provide this. Every species of mammal feeds its young on milk – even the primitive duck-billed platypus. Care of the young has been farthest developed among mammals, the education of human children being the most advanced example of this procedure.

Almost all the birds take care of their young. In many cases they build nests which are inaccessible to marauding enemies and later collect food for their young, both parents being usually responsible. The cuckoos are an exception to this rule, as the female cuckoo lays its egg in another bird's nest. When the young cuckoo hatches out it pushes the rightful occupants out of the nest. The foster parents – often smaller birds – feed and provide for it, with no seeming awareness of the fact that the monstrous interloper is not their own offspring.

Throughout the rest of the animal kingdom the general practice is to leave the young to fend for themselves. Eggs are often deposited with some regard either for safety or for future food supplies, but consideration seldom goes any farther. There are instances, however, of animals other than mammals and birds taking care of their young, but these are exceptions to the general practice. Perhaps the best known of these are the 'social' insects – those particular species of ants, bees and wasps which live in communities whose whole purpose is concerned with the rearing of the young. The common stickleback of ponds and canals is a fish which, exceptionally, cares for its young. What is even more odd in the case of the stickleback, is that it is the male fish that is responsible for building the nest in which the eggs are laid, and for protecting the young fry until they are able to take care of themselves.

The illustrations for Lesson 21 in the Pupils' Book show

1  the reproduction of the freshwater hydra from the parent
2  cat and kittens
3  an embryo chicken in its shell (other contents of the egg have been omitted in order that the embryo may be shown more clearly)
4  a lackey moth and eggs
5  plaice eggs and young, showing yolk sac
6  hen blackbird and young
7  colobus monkey and young.

## CODE

1  Observe any examples of the skeletal remains of a colony formed by animals which reproduce directly from the parent, *e.g.* coral, sponge, sea-mat.
   The skeleton remains of coral polyps and sponge may be retained as they are. Sea-mat is better preserved in 5% formaldehyde. It may also be pressed and mounted. Sea-mat is quite commonly found washed up on beaches and is often mistaken for a seaweed.

2  Observe any animal which is able to reproduce by dividing or 'budding', *e.g.* starfish, sea-anemone, hydra.
   If a specimen of a starfish developing from one ray is obtained, it may be kept as it is, or preserved in 5% formaldehyde.

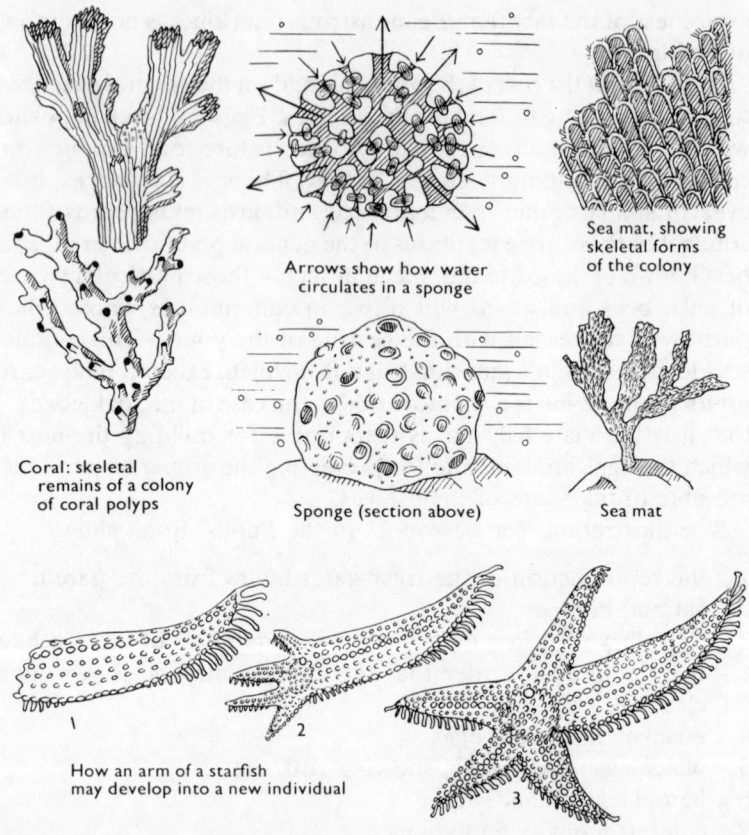

Arrows show how water
circulates in a sponge

Sea mat, showing
skeletal forms
of the colony

Coral: skeletal
remains of a colony
of coral polyps

Sponge (section above)

Sea mat

How an arm of a starfish
may develop into a new individual

Sea-anemones may be preserved in 5% formaldehyde.

Hydra are very small freshwater animals and are often collected
by accident amongst freshwater plants. They can be kept alive
without any difficulty in pond water in a glass container with
plenty of sunlight, where their reproduction by 'budding' can be
easily observed.

3  Observe any examples of eggs, or of egg cases found on the
shore.

The egg cases opposite may be preserved in 5% formaldehyde,
but they will keep for quite a long time without being preserved.

4  Observe some young animals which were born without being
contained in an egg, e.g. lambs, calves, or pet mammals such as
kittens, puppies, hamsters, rabbits, mice, guinea pigs.

5  Collect frogspawn for the lesson on amphibians.

182

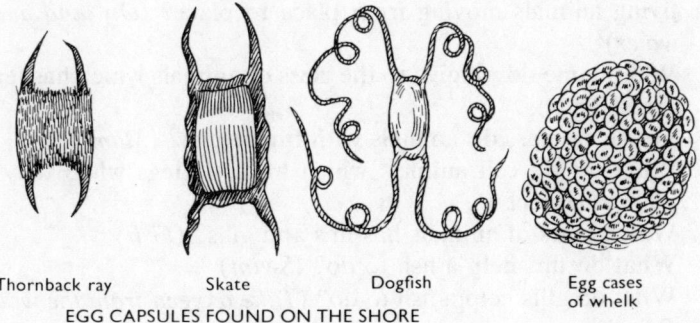

Thornback ray          Skate          Dogfish          Egg cases
                                                        of whelk
EGG CAPSULES FOUND ON THE SHORE

## Written Work

1  Some simple animals grow from part of the parent.
2  Most kinds of <u>animals</u> have young by laying <u>eggs</u>.
3  Most <u>mammals</u> are born as tiny <u>baby</u> animals.
4  <u>Mammals</u> and most birds take <u>care</u> of their young.

## (22) SOME WATER-AND-LAND ANIMALS

### AMPHIBIANS

## Demonstration Material

Any or all of the following:

1  Eggs of frog, toad or newt.
2  Tadpoles of frog, toad or newt.
3  Adult frog, toad or newt.

## Sample Link Questions

1  What are the four main needs of living things? (*Oxygen, food, to grow, to have young*)
2  Some living animals can fly in the air. Where else do we find

living animals moving from place to place? (*On land and in water*)

3  What name do we give to the class of animals which has feathers? (*Birds*)
4  To which class do animals with fur belong? (*Mammals*)
5  What do we call animals which have six legs when they are adults? (*Insects*)
6  Which class of animals has fins and gills? (*Fish*)
7  What do fins help a fish to do? (*Swim*)
8  What do gills help a fish to do? (*Take oxygen from the water – Breathe*)
9  What is it that enables some animals to take oxygen from the air? (*Lungs*)
10 What are the three main ways in which animals have young? (*Some young animals grow from part of the parent. Some are born from eggs. Some are born as tiny babies*)

## Relevant Information

In Pupils' Book 1 four classes of animals were introduced:

1  mammals – animals with hair (whose young drink milk)
2  birds – animals with feathers
3  insects – animals with six legs
4  fish – animals with fins and gills.

The main points covered by this lesson are:

1  Amphibians (Greek *amphi*, both; *bios*, life) are a class of animals which evolved from primitive fish, and as a class they live partly in fresh water and partly on the land.
2  The skins of present-day amphibians are either smooth and moist, or rough and dry, but never scaly like those of most fish, or those of reptiles.
3  Modern adult amphibians are of three main kinds:
   *a* those with four legs and a tail
   *b* those with four legs and no tail
   *c* those with no legs and no tail.
4  British amphibians are typical of the majority of amphibians, in that:
   *a* their eggs are laid in water

184

*b* their tadpoles feed in the water and breathe by means of gills
*c* the adults feed on land and breathe by means of lungs.

The general life history of the British amphibians, and the difference between newts, frogs and toads, are dealt with in Book 3.

Many animals – for example, the crocodile, the otter and the penguin – are equally at home on land and in water; they can be called *amphibious*. The true amphibians, however, are a class of animals which are mid-way in evolution between fish and reptiles.

Fossil evidence indicates that an ancient and now extinct group of primitive fish were ancestors of the lung-fishes, and also ancestors of the ancient amphibians.

The ancient amphibians began to appear some 300 million years ago and were the first of the backboned animals to walk on the land. They were not the first animals on land, as insects and other animals without backbones were already in evidence among the clubmosses, horsetails and ferns which grew in the coal-age swamps of that time.

The first of these ancient amphibians probably spent most of their time in the water, emerging to crawl through the mud from pool to pool when necessary. Their wriggling movements through the mud were probably assisted by the two sets of paired fins which fish normally have. As they became more adapted to moving, breathing and feeding on the land, they became quite distinct from the lung-

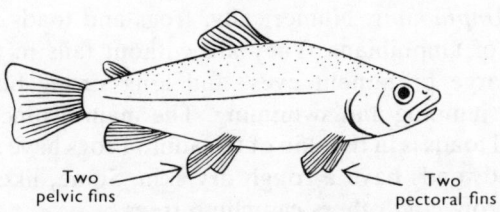

Two
pelvic fins

Two
pectoral fins

The paired fins of a fish, from which the four limbs
of the amphibians probably developed

fishes, in that the fins became bony limbs and the gills disappeared with the development of the lungs. Fossils show that at least some of these amphibians had, in the adult stage, long bodies, flattened tails, and some similarity in outline to the present-day salamanders and newts.

From the ancient amphibians sprang the ancestors of the modern amphibians. Although in the period of the coal-age swamps these amphibians became the dominant land animals, their place was gradually usurped by other backboned animals better adapted for a life on land.

## Modern Amphibians

Modern amphibians are backboned animals. They are cold-blooded, *i.e.* their temperatures vary according to the surroundings. Their skins are either smooth and moist, or rough and dry, but they have no visible scales. They belong to one of three groups:

1   the tailed amphibians – the salamanders and newts
2   the tailless amphibians – the frogs and toads
3   the legless amphibians – the caecilians.

*The Tailed Amphibians.* There is no essential difference between salamanders and newts, except for size in the adult phase. The larger species are salamanders, and the smaller species newts or efts. It is also accepted that, in general, newts have flattened tails to assist in swimming, whereas salamanders have tails which are more rounded and less compressed. Because of their appearance, they are sometimes mistakenly called lizards.

*The Tailless Amphibians.* Numerically, frogs and toads form the largest group of amphibians. They are without tails in the adult phase, have large prominent eyes, and long strong hind limbs developed for jumping and swimming. The main distinction between frogs and toads is in the skin of the adult. Frogs have a smooth moist skin, and toads have a rough dry skin. Some, like the pug toads, can burrow, and others can climb trees.

*The Legless Amphibians.* The caecilians are rare amphibians with neither limbs nor tails apparent. They are found in India, South Africa, Central and South America, where they are almost entirely burrowing animals. The adult caecilian has a cylindrical body with ring-like folds, and a small pointed head with sharp teeth. Very tiny scales are hidden in the folds of the skin – evidence of their fish ancestors. Because of their habits, the eyes of caecilians are virtually useless, and they are sometimes referred to as *blind-worms*. In appearance a caecilian does resemble a large blue-grey earthworm.

## The General Characteristics of Modern Amphibians

The typical modern amphibian has a life divided into three distinct stages – egg, tadpole and adult. The tadpole which hatches out of the egg is a freshwater animal, feeding in the water, and with gills to obtain its oxygen from the water. In the adult stage, it is a land animal living in moist surroundings, and generally obtaining its food from the land. It obtains a considerable amount of its oxygen by absorption through the skin, but it breathes by means of lungs in addition.

Modern amphibians have good ears. On the common frog these look like dark patches of skin behind the eyes and may easily be examined. It was in the amphibian class that tongues first made their appearance, as a developed muscle on the floor of the mouth.

Glands on the backs of modern amphibians can excrete defensive fluids which make the animals unpalatable to enemies. The warts on the back of a toad are accumulations of these glands. The secretion from the skin of the South American poison frogs is particularly venomous and has been used by certain tribes of Indians during the preparation of poisoned arrows. Generally speaking, however, the secretion from an amphibian's skin is of no harm to humans and is poisonous only if taken internally. Children should wash their hands after handling amphibians.

As a class, the amphibians are useful to human beings, because of the large quantity of worms and insects which they consume, to say nothing of other small animals. No matter how repulsive they appear, they should not be killed but should be welcomed.

Although prehistoric amphibians reached a fairly large size, the majority of modern amphibians are small animals, relatively speaking. The largest British amphibian is the great crested or warty newt which attains some 15–18 centimetres in length. The largest frog in the world is the goliath frog of West Africa. This is a rare animal, reaching some 30 centimetres in length. The largest of all modern amphibians is the giant salamander of Japan, which may grow to over 1.5 metres in length.

Some exceptional species do not have all the characteristics of amphibians. Differences are usually associated with methods of obtaining oxygen or with methods of having young. The majority of amphibians have young by means of eggs, but there are some species which have their eggs hatched out before birth. The spotted salamander of central and southern Europe, for example, retains its eggs until after they have hatched out and developed limbs. Then the mother enters the water where the young are born as free-swimming tadpoles. Most of the amphibians which have young by means of eggs lay them in the water, but some species lay them on the land. Some caecilians burrow chambers in the mud, and there the tadpoles hatch, making their own way to the water. There are also amphibians which care for their young, or at least for their eggs. Some frogs and toads – for example, the male midwife toad – carry their eggs about until they are ready to hatch. The eggs of the Surinam toad become embedded in tiny pits in the back of the mother, where they remain until the tadpole stage ends and the young toads are ready to emerge.

But just as some amphibians seem able to spend all their time on the land, so others seem able to spend it in fresh water. The North American axolotl is a salamander which can reach sexual maturity while still in the gilled stage, and can therefore reproduce without ever leaving the water. It may, on the other hand, metamorphose, lose its gills, develop lungs, and emerge on to the land to be known from then on as a lung-breathing salamander.

Such exceptions show that the amphibians are a transitional class in the trend of evolution, straddled awkwardly across the barrier which separates the animals with a complete life in water from the animals with a complete life on the land.

Despite these exceptions, the majority of the amphibians emerge as tadpoles from eggs laid in water. They spend the first phase of their lives feeding in water and breathing by means of gills, and they eventually metamorphose into animals with lungs which can live and feed on land, but which have to return to water to lay their eggs. This applies to all the British amphibians.

## Amphibians in Britain

No caecilians are found in the British Isles, but representatives of both tailed and tailless amphibians are found in the form of newts, frogs and toads. The eggs of all three are laid in fresh water during spring. From these eggs hatch tadpoles (sometimes referred to as larvae). The growing tadpoles feed in the water and obtain their oxygen by means of gills. When finally the lungs have developed and the gills have withered, the young amphibians are able to leave the water and feed on other small animals living on land. In the adult stage they absorb oxygen through the skin, and also by forcing air into the lungs. As they are cold-blooded, they hibernate through winter. In the adult stage, British amphibians are easily distinguished from one another.

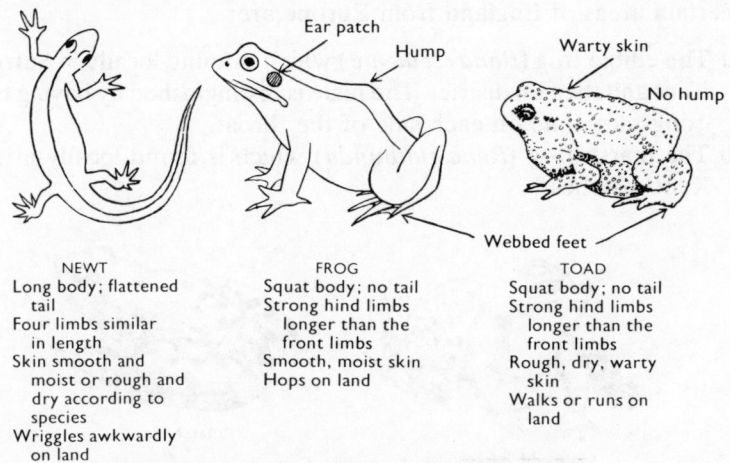

| NEWT | FROG | TOAD |
|---|---|---|
| Long body; flattened tail | Squat body; no tail | Squat body; no tail |
| Four limbs similar in length | Strong hind limbs longer than the front limbs | Strong hind limbs longer than the front limbs |
| Skin smooth and moist or rough and dry according to species | Smooth, moist skin | Rough, dry, warty skin |
| Wriggles awkwardly on land | Hops on land | Walks or runs on land |

ADULT AMPHIBIANS FOUND IN THE BRITISH ISLES

## The British Frogs

In the adult stage frogs have squat bodies and are without tails. The hind legs are stronger and much longer than the front legs. There are four clawless fingers on each front foot, and five clawless toes on each rear foot. The toes on the rear feet are webbed to assist in swimming, as frogs use their rear feet to force themselves through water. A severed limb can be regrown in the tadpole stage but not in the adult stage. Adults are able to adapt their colour to a certain extent to the surroundings. That is why common frogs are sometimes greenish and sometimes brownish.

The skin of a British frog is smooth and moist, as distinct from that of a toad which appears rough, dry and warty. On land it uses its powerful hind limbs to leap from place to place.

During spring and summer frogs are found in cool damp places, often amongst grass and in or near fresh water. They feed on insects, worms and other small animals.

From about October to early spring they hibernate, usually in muddy holes or in the mud at the bottom of ponds and ditches, where their reduced oxygen requirements are met by respiration through the skin.

189

The common frog (*Rana temporaria*) is the only frog native to the British Isles. It is sometimes called the 'grass frog', and is the one most commonly found by children. It is very widely distributed.

Two other species of frogs which have been introduced into certain areas of England from Europe are:

*a* The edible frog (*Rana esculenta*) which is found locally in Surrey, Kent and the Fen district. The male is distinguished by having two 'song sacs', one on each side of the throat.

*b* The marsh frog (*Rana ridibunda*) which is found locally in the Kent marshes.

Male edible frog with
' song sacs ' inflated

Marsh frog

### The British Toads

Adult toads are similar to frogs in having squat tailless bodies with hind legs stronger and longer than the fore-legs. Like frogs, toads have four clawless fingers on each front foot, and five clawless toes on each rear foot. Also like frogs, toads have their rear feet webbed to assist in swimming. In general, the hind legs are shorter than those of a frog and the body is broader. Toads can also change colour to a certain extent according to their surroundings.

The rough, dry skin of British toads distinguishes them from frogs. Even at a distance, however, they may be distinguished, for whereas a frog leaps, a toad walks or runs (although it can jump if necessary).

Toads tend to be creatures of habit. During the day they hide, usually in some cool damp hole, although they may sometimes be seen walking about when it is raining. They hunt for their food at night, returning to their hideouts before dawn. At night a torch will sometimes reveal one sitting in a shallow pool. Because of the quantities of small animals which they eat, they are invaluable allies to the gardener.

During winter they hibernate, often in a deep hole, well away from water. They take five years to reach maturity and can live longer than twelve years.

Two species of toads are native to the British Isles:

a The common toad (*Bufo bufo*). This is the toad most commonly found. It is very widely distributed, but not in Ireland. The female is about 10 centimetres long, and the male about 7 centimetres long. When surprised, the common toad has a tendency to raise itself on all four limbs and turn its back. This is due to alarm and not disgust.

b The natterjack toad (*Bufo calamita*). This is distributed throughout the British Isles, but not as widely as the common toad. The natterjack is easily distinguished from the common toad by the yellow line which runs down the centre of its back. It is also smaller than the common toad, both sexes being about 6 centimetres long. It is sometimes called the 'running toad', as it runs, whereas the common toad more often walks. The bell or midwife toad has been introduced into some parts of Britain. The male midwife toad carries the string of eggs attached to its rear feet until just before hatching.

Natterjack toad

Male midwife toad
carrying eggs

## The British Newts

The tails of newts are flattened and retained throughout life. A newt uses its tail in the water for swimming, the limbs being of little consequence. It moves swiftly in water, but on land its movements are awkward and ungainly, suggestive of the difficulties encountered by its early ancestors. As on the frog and toad, only four clawless fingers are visible on the front limbs, but on the rear limbs there are five clawless toes. If part of a limb is lost, it can be regrown from the stump.

During spring and summer newts are usually found in water where they may remain until autumn. In the water they feed on small aquatic animals. When on land, they hide in damp places during the day, emerging at night to feed on such small animals as insects, centipedes and small worms.

During winter they hibernate in damp underground places, often under stones, where numbers of them are sometimes found twisted together.

There are three species of newts native to the British Isles:

*a* The common or smooth newt (*Triturus cristatus*). These are the newts most commonly found in ponds, canals and ditches in spring. During the breeding season the male has a continuous wavy crest running from head to tail. The belly is orange and dappled. Its beauty is far greater than that of the female, whose olive-brown body lacks a crest. In fact the female is so mediocre by comparison, that it is sometimes assumed to be a different species. Both adults are about 10 centimetres long.

*b* The crested newt (*Triturus palustris*). These are the largest of the British newts with a body length of about 15 to 18 centimetres. Whereas the common and palmate newts have smooth moist skins, the crested newt has a rough dry warty skin, similar to that of a toad. During the breeding season the male has a large saw-toothed crest, the line of which is broken above the hips. It is also known as the 'great crested newt' or as the 'warty newt'.

*c* The palmate newt (*Triturus helveticus*). These are the smallest of the British newts; with a body length of about 7 centimetres. The tip of the tail is thread-like, and is longer on the male than on the

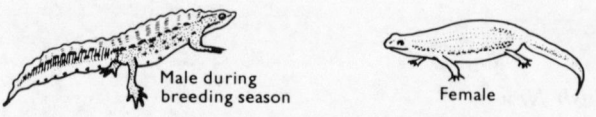

Male during breeding season          Female

THE COMMON OR SMOOTH NEWT

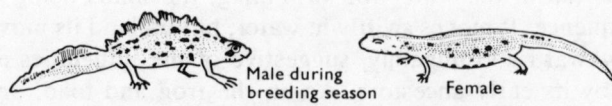

Male during breeding season          Female

THE CRESTED OR WARTY NEWT

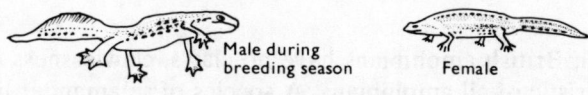

Male during
breeding season

Female

PALMATE NEWTS

female. During the breeding season the male develops a very low, unbroken crest which is not very noticeable. It also develops slight black webbing between the toes.

The crests of the male newts are developed for the breeding season only. During winter they are lost.

Children sometimes confuse newts with lizards. A newt is an *amphibian*, and a lizard is a *reptile*, therefore there are quite a number of differences. The most easily observable differences are:

| | |
|---|---|
| 1 A newt's tail is slightly flattened. | A lizard's tail is long and rounded. |
| 2 A newt has a skin which is either smooth and moist, or rough and dry, but never scaly. | A lizard has a scaly skin. |
| 3 A newt has no claws on its fingers and toes. | A lizard has claws on its fingers and toes. |

On land, of course, a newt's movements are slow and ungainly, whereas those of a lizard can be very swift. Thus, although children could catch a newt quite easily on land, they would probably have extreme difficulty in catching a lizard.

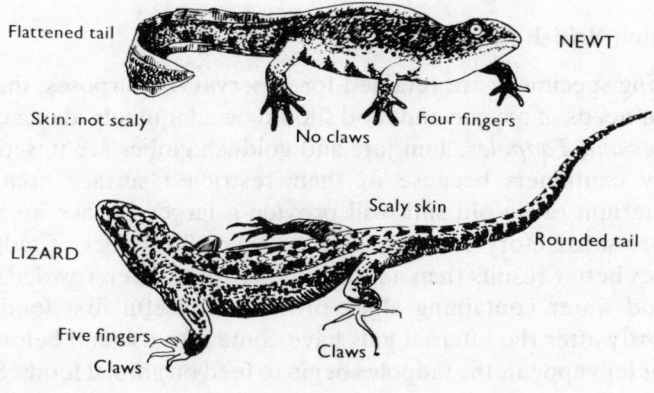

Flattened tail

NEWT

Skin: not scaly

No claws

Four fingers

Scaly skin

LIZARD

Rounded tail

Five fingers

Claws

Claws

*Note*

Although British amphibians have no claws, clawlessness is not a characteristic of all amphibians. A species of salamander living in the mountain streams of Japan has developed claws on its fingers and toes to assist it in gripping rocks when crawling against rushing water.

*Final Notes*

1  Most amphibians have young by means of eggs, but some do not.
2  Most of the amphibians which have young by means of eggs lay their eggs in water, but there are some which do not.
3  Most amphibians spend their tadpole stage in water, but some do not.
4  Most adult amphibians exist for the greater part of their lives on land, but some do not. Some never even leave the water.
5  Most amphibians breathe by means of gills during the tadpole stage, but some do not. This applies particularly to those amphibians which spend the tadpole stage in an egg out of water.
6  Most adult amphibians supplement their oxygen supply by breathing air into lungs, but some do not. Some of the salamanders which have returned to an aquatic existence have lost the use of their lungs and depend entirely on the absorption of oxygen through the skin.
7  Amphibians are not the only land animals to lay eggs in water. Certain insects, *e.g.* the mosquito and dragonfly, also do this.

**Keeping British Amphibians**

If living specimens are retained for observation purposes, the two major needs of oxygen and food should be adequately allowed for.

*a Eggs and Tadpoles*. Jam jars and goldfish globes are unsatisfactory containers because of their restricted surface area. An aquarium or an old sink will provide a larger surface area and more satisfactory accommodation. A small number of tadpoles gives better results than a large number in an overcrowded tank. Pond water containing algae provides a useful first food, but shortly after the internal gills have come into use and before the rear legs appear, the tadpoles begin to feed on animal foods. Small pieces of raw meat suspended on a cotton thread will suffice, provided they are removed after a few hours to prevent pollution of the water.

b *Adult Amphibians.* The essentials in any container for adult amphibians are a pool, terrestrial conditions provided with shade, and an adequate food supply. For adult frogs and toads, earthworms and other small animals such as insects, woodlice and centipedes are suitable food. For adult newts, small earthworms, white worms, insects and centipedes are suitable.

FROGS' EGGS laid at the bottom of a pond and fertilised there, usually between February and May. Subsequently rise due to swelling of outer coating of jelly, and float in a mass at the surface.

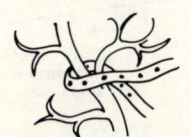

TOADS' EGGS Laid in a continuous string usually in April or May, wound round underwater plants whilst being fertilised.

NEWT EGGS Separate, already fertilised eggs are attached to the leaves of freshwater herbs, usually in April to late July. Not easily found.

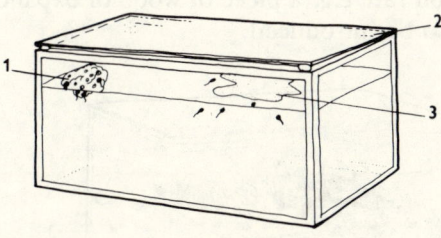

**Container for Eggs and Young Tadpoles**

1   Eggs and young tadpoles.
2   Glass cover raised on small blobs of plasticine. This restricts settlement of dust while allowing fresh air to pass over water surface.
3   Floating alga – the green 'scum' found in ponds. This is a suitable first food. Frog and toad tadpoles become carnivorous about the time that the front legs are beginning to appear. This is the time to introduce a daily feed of a small piece of raw meat

suspended on a cotton thread. It needs only to be about the size of a finger end, and should be removed after a few hours, to prevent pollution of the water.

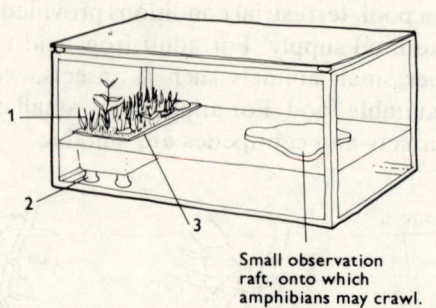

Small observation raft, onto which amphibians may crawl.

## Transitional Container for Tadpoles preparing to leave the Water

1 Waterproof container (*e.g.* plastic box) with living plants, *e.g.* turf, and small living animals. A section of stem infested with aphids, *e.g.* greenflies, adds to the food supply.
2 Small pots for raising container to suitable height.
3 Water up to rim of container so that tiny frogs can crawl out. A small observation raft, *e.g.* a piece of wood or expanded polystyrene may also be introduced.

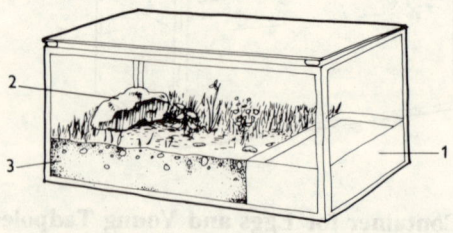

### Container for Adult Frogs and Toads

1 Pool. A plastic box or tin filled with water. Tap water may be used and changed occasionally.
2 Rock or piece of bark for shade.
3 Soil, turf and other living plants which should be watered occasionally. Earthworms and other small terrestrial animals, *e.g.* woodlice and centipedes, slugs and insects, may be introduced to the land area at intervals.

196

*Note*

If an aquarium is not available, a wooden box with a polythene lining (to restrict rotting of the wood below soil level) may be substituted.

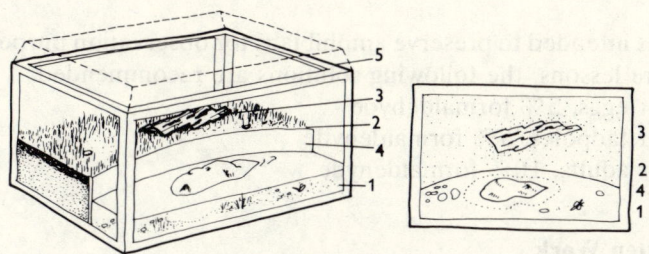

### A Container for Adult Newts

1  Water (with gravel and aquatic plants if desired).
2  Peat, topped by growing moss or grass, to which may be added very small worms plus insects, woodlice, and centipedes.
3  Rock or piece of bark for shade.
4  Curved strip of thick acetate sheeting or plastic to retain the peat.
5  Electric light shade.

*Note*

If no electric light shade is available, then the top should at least be covered so as to prevent the inmates escaping. Dissatisfied newts have a habit of climbing vertical surfaces and disappearing through small apertures, to be later discovered, dehydrated and dead in a dusty corner.

## CODE

1  Observe the egg, tadpole and adult stage of any amphibian.
2  Observe the gills (external) on any amphibian tadpole, and note that there are no gills on the adult.
3  Observe that an adult newt has a tail, but that an adult frog and an adult toad have no tails.
4  Examine the skin of any adult amphibian, and observe that
   *a* it is naked, *i.e.* it has no scales, feathers or hairs;
   *b* its skin is either

197

   *i* smooth and moist, as on the British frogs and the smooth
      and palmate newts
*or ii* rough and dry, as on the British toads and the great
      crested newt.

*Note*
If it is intended to preserve amphibians for observation purposes in
future lessons, the following solutions are recommended.
*a* For eggs, 5% formaldehyde
*b* For tadpoles, 5% formaldehyde
*c* For adults, 10% formaldehyde


**Written Work**

1   Early <u>fish</u> were the ancestors of the first <u>amphibians.</u>
2   Modern amphibians have no <u>scales.</u> Some have smooth <u>moist</u>
    skins. Others have <u>rough</u> dry skins.
3   <u>Salamanders</u> and <u>newts</u> have both legs and tails.
4   Frogs and toads have no <u>tails.</u> <u>Caecilians</u> have no legs.
5   British frogs, toads and newts lay <u>eggs</u> in <u>water.</u>
6   The tadpoles have <u>gills</u> to take oxygen from the water.
7   The adults have <u>lungs</u> to take oxygen from the air.

 **23**   **ANIMALS WITH SCALY SKINS
                AND LUNGS**

REPTILES

**Demonstration Material**
Any of the following:
1   A reptile, *e.g.*
    *a* tortoise or terrapin
    *b* lizard, alive or preserved
    *c* grass snake, alive or preserved.
2   Section of a reptile skin, showing scales, *e.g.* crocodile skin or
    snake skin.

198

## Sample Link Questions

1 What do we call animals with fins and gills? (*Fish*)
2 Which class of animals developed from fish? (*Amphibians*)
3 Which two kinds of amphibians have four legs and a tail when they are adults? (*Salamanders and newts*)
4 Which two kinds of amphibians have four legs but no tail when they are adults? (*Frogs and toads*)
5 Which amphibians have no legs and no tail when they are adults? (*Caecilians*)
6 What are the two kinds of skins to be found on adult amphibians? (*Smooth moist skins. Rough dry skins*)
7 Where do most adult amphibians feed? (*On land*)
8 Where do British amphibians lay their eggs? (*In fresh water*)
9 How do the tadpoles of British amphibians take oxygen from the water? (*Through gills*)
10 What do adult British amphibians use to help take oxygen from the air? (*Lungs*)


## Relevant Information

The main points covered by this lesson are:

1 reptiles are a class of animals which evolved from primitive amphibians, and, as a class, they are adapted from birth to withstand conditions on land;
2 the prehistoric monsters were reptiles;
3 modern reptiles have scaly skins, and their eggs are always laid on land;
4 they include the crocodiles, turtles, snakes and lizards.

Two of the main developments in the evolution of the amphibian class were:

1 lungs for taking oxygen from the air;
2 legs for moving from place to place on land.

Amphibians were unable to stray far from water, however, because of their relatively soft body skins and the jelly-like coverings of their eggs. Neither was capable in dry conditions of preventing death by loss of moisture. The further progress of evolution produced a higher class of animals capable of resisting dry conditions on land by the development of a thicker body skin, and eggs protected by a

199

tough shell which enabled them to survive when laid on land. These animals were the reptiles. Their four main needs could be satisfied on land. They had more efficient lungs than amphibians; they could obtain their food from things living on the land; and they could grow from egg to adult without having to return to water to reproduce. Of the backboned animals, they were the first true dwellers on land.

## The Age of Reptiles

It is probable that the first primitive reptiles evolved from the amphibians during the coal age. The era which followed the coal age saw the evolution of many families of reptiles. Amongst these were the ancestors of the modern reptiles, the ancestors of mammals, and the ancestors of birds. This era – often called the age of reptiles – began about 275 million years ago and lasted some 200 million years. During this age, while the coniferous plants were rising to dominate the plant kingdom, there lived and died some of the most fantastic animals ever to evolve. Amongst these were the dinosaurs.

## The Dinosaurs

Some of these reptiles were as small as a hen, but others attained a gigantic size, and they included the largest animals ever to walk on land. The word *reptile* means creeping or crawling, but many of the dinosaurs – in particular the flesheaters – developed the ability to walk and run on their hind legs. The plant-eaters generally walked on four feet. The most ferocious were the flesh-eating tyrannosaurus and allosaurus. The largest were the herbivorous diplodocus, brontosaurus and brachiosaurus. In spite of their huge size, the herbivores were stupid and defenceless against the flesh-eaters, but later there evolved armoured herbivores such as the stegosaurus and triceratops which had a certain protection.

During the first 25 to 30 million years of their development, the dinosaurs reached their peak, and their distribution became world-wide. Eventually their rule began to wane, and the last of them became extinct some 60 to 70 million years ago. This was long before the days of cave men; as primitive man did not appear on the scene until well within the last million years, he arrived far too late to enjoy their company.

# REPTILES

## DINOSAURS

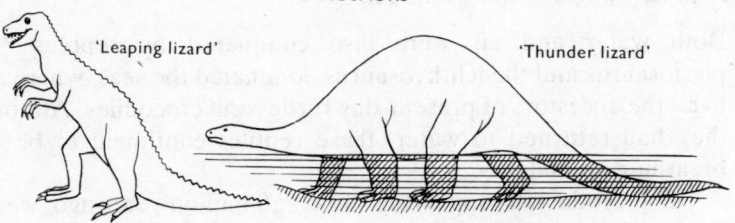

'Leaping lizard'

'Thunder lizard'

**ALLOSAURUS**
Flesh-eating monster, over
9m long. Sharp claws and
sabre teeth. Lived about 170
million years ago

**BRONTOSAURUS**
Plant-eating monster, over 21m
long. Weight about 35 tonnes. Often
buoyed its body up with water. Lived
about 160 million years ago. The
diplodocus was similar but 6m
longer

**STEGOSAURUS**
Plant-eating monster, about
7m long. Two rows of
jutting bony plates. Lived
about 140 million years ago

**TRICERATOPS**
Plant-eating monster, about
6m long, and with three
horns. Lived about 100
million years ago

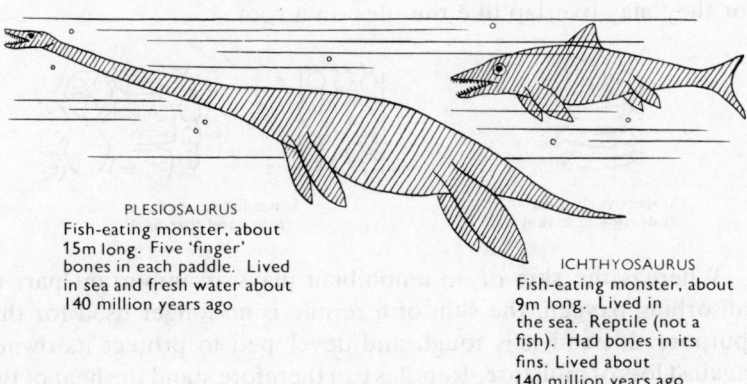

**PLESIOSAURUS**
Fish-eating monster, about
15m long. Five 'finger'
bones in each paddle. Lived
in sea and fresh water about
140 million years ago

**ICHTHYOSAURUS**
Fish-eating monster, about
9m long. Lived in
the sea. Reptile (not a
fish). Had bones in its
fins. Lived about
140 million years ago

*Reptiles in Water and in the Air*

Both water and air were also conquered by reptiles. The plesiosaurus and the ichthyosaurus dominated the seas, where also lived the ancestors of present-day turtles and crocodiles. Although they had returned to water, these reptiles continued to be air-breathing animals.

The pterodactyls, appearing some 150 million years ago, were a family of reptiles which developed the ability to fly. Each of their fore limbs was strong, with one finger developed into an elongated support from which a membrane (like a bat's wing) extended to the thigh. Size varied from that of a sparrow to that of the largest – the pteranodon, which had a wing span of up to 8 metres. Although birds evolved from reptiles, pterodactyls are not credited with being their ancestors. It is also unlikely that pterodactyls had anything like the flying ability of birds.

Knowledge of the various periods in which these animals lived is based on fossil evidence, and therefore stated times are approximate. With the extinction of the dinosaurs and many other reptiles, the dominance of the animal kingdom passed to the birds and the mammals. In modern times only four groups of reptiles exist.

## Modern Reptiles

Like the amphibians, modern reptiles are cold-blooded backboned animals. The skin of a reptile, however, is noticeably different from that of an amphibian; not only is it dry but it is covered with small horny plates which are called scales. These scales differ in structure from those of a fish. They may be flat, fitting together like a mosaic, or they may overlap like the tiles on a roof.

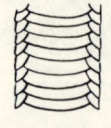 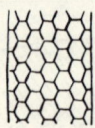 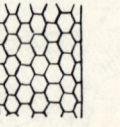 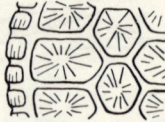

Scales overlapping
(underside of snake)

Scales fitting together
(snake and tortoise)

Whereas the skin of an amphibian plays an important part in absorbing oxygen, the skin of a reptile is no longer used for this purpose. Instead it is tough and developed to protect its owner against loss of moisture. Reptiles can therefore stand the heat of the sun and can live in very dry surroundings. Many of the reptiles slough their skins periodically as they grow, thus exposing new clean layers beneath.

Unlike the amphibian egg, which is surrounded only by jelly, the reptile egg is protected by a tough covering and is always laid on land, where in many cases it is hatched by the heat of the sun. The egg itself is comparatively large and contains a quantity of food yolk. The tough covering is usually parchment-like, but some are hard and limey like the shell of a bird's egg. The colour may be white or yellowish but is never speckled in the way some birds' eggs are. Some reptiles retain their eggs until hatched, the young being born as tiny replicas of the parent.

Throughout their lives all reptiles breathe by means of lungs. There is no tadpole stage with gills. All modern reptiles have tongues. They belong to one of the following groups.

1　The tuatara
2　The crocodiles and alligators – including the caymans
3　The turtles – including the tortoises and terrapins
4　The lizards and snakes

1 *The tuatara* is a rare primitive reptile found on the islands of Cook Strait, New Zealand.

2 *The crocodiles, alligators and caymans* have changed very little from their prehistoric ancestors. They use their long powerful tails for swimming, and when on land they spend their time basking in the sun. Food consists of other living animals which they catch in the water. Their oval eggs, which are laid on land, are about the size of a hen's egg and covered by a limey shell. These animals hunt in fresh water, except for the salt-water crocodile, which is the commonest and largest of all modern reptiles, reaching 6 to 9 metres in length.

There is little external difference between crocodiles and alligators. On crocodiles, the fourth tooth of the lower jaw fits into a noticeable notch in the side of the upper jaw. (See illustration in Pupils' Book 2.) The alligators and caymans have this tooth fitting into a concealed pit in the upper jaw.

3 *The turtles, tortoises and terrapins* are little changed from their primitive ancestors. Their 'shell' consists of a bony structure in two halves; the upper half is the carapace, and the lower half is the

plastron. The horny plates of the shell are scales, which fit together but are not shed periodically like those elsewhere on the body.

This protective outer covering varies according to species. The land turtles have a high domed shape to give full protection to head and body, while the water turtles have a flattened shape which is too small to accommodate the head and limbs. The eggs, which are always laid on land, vary according to species; some are covered by a parchment-like layer, and others have a hard limey covering. They are buried in the ground, after which parental interest ceases. Although terminology is far from specific, the name *turtle* is given to those which live in the sea, *tortoise* to those which live on land, and *terrapin* to those which live in fresh water. In general, those living in salt water have their fingers and toes fused so that their limbs are like definite paddles, while those living in fresh water have their feet webbed to assist in swimming. Again generally, those which live in water feed on animal foods, while those which live on land feed on plant foods. The largest of these reptiles is the Pacific leatherback turtle with an overall length of about 2 metres. Tortoises are also the longest lived of reptiles, and indeed of all the vertebrate animals. A Mediterranean spur-thighed tortoise is known to have lived for over 116 years, but there are other reasonably reliable records of tortoises living up to 140 years.

4 *Lizards and snakes* are closely related and are the most successfully established of modern reptiles. They are most common in tropical regions.

*Lizards.* The 2500 species of lizards in the world include the skinks, chameleons, monitors, iguanas, geckoes, glass snakes, and slow-worms. The majority have two pairs of well-developed limbs with five fingers or toes to each foot. There are some, however, like the slow-worms, which are without visible signs of limbs, and are thus snake-like in appearance. Most lizards feed on insects, but some feed on other animals and some on plant food. Although some lizards' eggs are hard-shelled, most have soft shells, and are laid in burrows or buried in the ground. Some, like the common British lizard, retain them until the young have hatched, so that they are born free of the egg.

A lizard's tail serves as a form of defence, as it will become detached if seized by an enemy. The lizard then grows another.

Methods of movement vary, from the frilled lizard which can run on its hind legs, suggestive of the long-dead dinosaurs, to the so-called flying dragon of Indonesia which can glide from tree to tree by means of skin flaps. Unlike snakes, lizards close their eyes when sleeping, and on some an external ear patch is visible, a feature which snakes lack. Only two are poisonous – the gila monster, and beaded lizard of Texas, Arizona and Mexico. The largest lizard in the world is the monitor, the komodo dragon, with a length of 2 to 3 metres, and a weight of about 70–90 kilograms.

*Snakes* include in their 2000 species the boas, pythons, vipers and adders. Most are without visible limbs, but a few show clawlike traces of the two rear limbs. All feed on animal foods such as insects, mammals, birds, fish, amphibians, and other reptiles. Prey may be swallowed alive, or killed first by crushing or by poisoning. Boas, for example, crush their victims, and vipers poison them. A snake positions its victim so that it is swallowed head first – it goes down more easily that way. The two halves of a snake's lower jaw are joined by an elastic ligament which permits it to swallow animals larger than the size of its own head. In this respect it differs from a lizard. During swallowing, the snake's windpipe is pushed out of its mouth so that it can continue breathing. Teeth are hooked to assist in swallowing, and in the poisonous snakes, certain teeth in the upper jar are modified into poison fangs. The tongue is forked and slender, serving as a sense organ to investigate food before swallowing.

Most snakes have young by means of eggs, but many vipers retain them until after the young have hatched. The Indian cobra and some boas guard their eggs by enclosing them within the coils of the body. The longest of modern snakes are the reticulated python of the Far East with a possible length of over 9 metres, and the anaconda of tropical South America with a possible length of over 11 metres.

## The British Reptiles

These are the sand lizard, common lizard, slow-worm, grass snake, smooth, snake, and the viper or adder (our only poisonous snake). During the winter British reptiles hibernate, but even when they are not in hibernation, they are seldom encountered. Tortoises and terrapins kept as pets are, of course, imported animals.

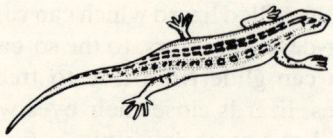

**SAND LIZARD.** Confined almost to Dorset, Surrey, and sandhills along Lancashire coast. Head broader than that of common lizard. Female looks purple-brown, but male is greenish.

**COMMON LIZARD.** Widely distributed throughout British Isles, particularly on heaths and amongst heather. Colour brownish. Runs swiftly and jumps. Two ear patches are visible.

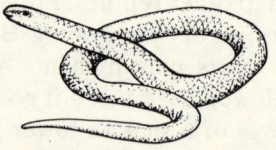

**SLOW WORM** also known as 'blind-worm' and 'deaf adder', but it is not slow, blind, deaf, nor an adder. Common throughout British Isles, excluding Ireland, on heaths and woodland fringes. Although legless, it is distinguished from British snakes by the thin dark line down the centre of its back. If the tip of the tail is rounded instead of pointed, this indicates that a new length is growing to replace one that has been lost.

**GRASS SNAKE.** Widely distributed throughout British Isles excluding Ireland. Easily distinguished by bright yellow, orange or white collar, behind which are two black patches.

SMOOTH SNAKE. Least common of British snakes, and confined to a few southern counties, particularly where the sand lizard (upon which it will prey) occurs. Ground colour on the upper surface is grey, brown, or reddish. Like the grass snake, it emits an objectionable smell when captured, but also like the grass snake, is easily tamed.

VIPER OR ADDER is the only poisonous British reptile. Widespread through England, Wales and Scotland, usually on commons, sandy heaths and hedgebanks. Distinguished by black zig-zag line running centrally down the back, and a V-shaped mark on the head. Eyes are a bright coppery red with vertical slit-like pupils. Not aggressive to humans, but will bite in self-defence if trodden on.

## Keeping Tortoises

Perhaps the commonest reptiles kept as pets are tortoises. Being reptiles, they like sunshine. It is not advisable to keep them in a school, unless they can be provided with suitable outdoor accommodation during the summer months. If it is desired to keep tortoises, and outdoor facilities are available, the following points are worthy of note.

1 *Accommodation*. A restricted area of freedom, *e.g.* a wire-netting enclosure, should include access to:
   *a* dry weather-proof quarters
   *b* a general foraging and exercising area, *e.g.* a rockery with small shrubs and herbs
   *c* a shallow sunken pool of water suitable for bathing and drinking purposes. This should be shaded from the full heat of the sun, and, if large enough, could also be used for terrapins.
2 *Food*. Tortoises are mainly plant-eaters. Suitable foods include lettuce leaves, tomato, dandelions, peas, etc., and even grass. Bread is sometimes eaten.
3 *Hibernation*. It is better to allow tortoises to hibernate completely during the winter months. A box filled with straw or torn newspapers is a suitable container. This should be placed in a cool dry place, *e.g.* a shed, but not where the temperature is liable to drop below freezing point.
4 *Purchase*. The tortoises sold in pet shops in this country are

usually the so-called 'Greek' tortoises imported from countries bordering on the Mediterranean. The treatment they receive during importation is often unsatisfactory from the point of view of a tortoise. By the time they reach the customer, they are often under-fed, and not infrequently damaged, although in recent years, measures have been taken to combat these faults. If tortoises are to be purchased, the following points are worthy of observation:

a Small young tortoises are not as hardy as adults. Specimens exceeding 12 centimetres in length are usually more suitable.

b May, June and July are usually the best months for purchasing.

c The eyes should be bright, and without any matter.

d The specimens selected should be the heaviest possible for their size.

e They should also be the liveliest.

f The shell should be undamaged.

g The hind quarters should be free from body lice. The body louse is a small eight-legged animal which feeds on the tortoise by burying its head in the reptile's skin. It is sometimes mistakenly called a baby tortoise! Care should be taken when removing a body louse, as there is a danger of its head breaking off and remaining in the flesh of the tortoise. If one is discovered, it may be removed in the following way:

    i kill the parasite first, e.g. smother with cotton wool soaked in methylated spirits, paraffin, or strong dettol;

    ii when it is dead and relaxed, remove gently with tweezers;

    iii paint the wound with an antiseptic, e.g. dettol solution or penicillin ointment.

*Notes*

1 Freshly-purchased tortoises may not feed readily at first due to the treatment they have experienced. They may be encouraged to feed by keeping for a while in a warm temperature, e.g. summer temperatures of 25°Celsius. They will feed very little at temperatures below 15°Celsius.

2 The age of a tortoise is difficult to judge. The belief that each ring on the carapace represents a year of growth is false, as a baby tortoise may grow several rings in one year.

3 For those children who inevitably wish to know whether a

tortoise is male or female, the following generalisations may be helpful:

a  On a male, the lower shell (plastron) is usually curved inwards near the tail. On a female, it is usually flat.

b  On a male, the tail is usually long and pointed. On a female, the tail is usually short and fat.

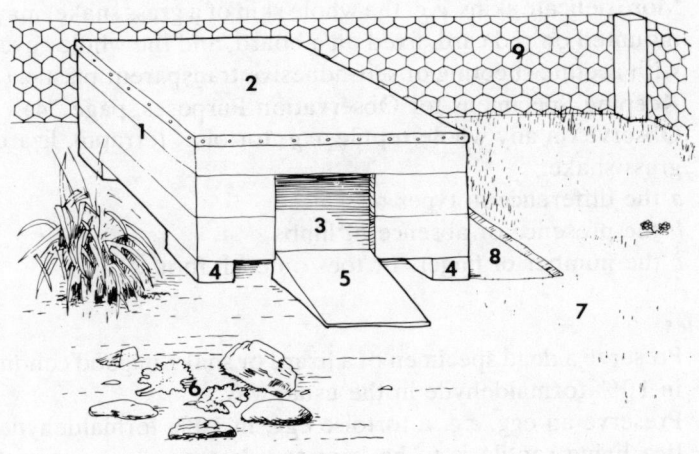

## Suggested Outdoor Accommodation for Tortoises

1  Sleeping quarters, *e.g.* a painted wooden box.
2  Waterproof covering, *e.g.* a strip of tarpaulin.
3  Entrance – preferably facing east to receive the rays of the morning sun.
4  Supporting bricks – to raise the box off the ground.
5  Access ramp.
6  General foraging area and food.
7  'Lawn'.
8  Shallow sunken pool – in this case on the north side of the box, so as to be shaded from full sunlight.
9  Wire-netting enclosure – say 45 centimetres high, with its lower edge buried in the ground to prevent escape by burrowing. Boring a hole in the shell of the tortoise and fastening it to a stake by a length of string is not a satisfactory method of restricting its movements.

## CODE

1 Observe the scales on the skin of any reptile. Notice whether they fit together as on the shell of a tortoise, or overlap as on the stomach of a snake. Sections of skin not liable to damage by handling may be retained as they are. Examples are crocodile skin and snakeskin of the kind used in the manufacture of various leather goods.

More delicate skins, *e.g.* the whole skin of a grass snake, may be mounted on red card, fixed on a board, and the whole covered with acetate sheeting or self-adhesive transparent plastic. (See Keeping Specimens for Observation Purposes, page 24).

2 Observe on any whole reptile, *e.g.* tortoise, terrapin, lizard or grass snake:

*a* the difference in types of scales

*b* the presence or absence of limbs

*c* the number of fingers or toes on each foot.

*Notes*

1 Preserve a dead specimen of a lizard or snake in good condition in 10% formaldehyde in the usual way.

2 Preserve an egg, *e.g.* a tortoise egg, in 10% formaldehyde.

3 If a living reptile is to be kept for classroom purposes, then correct food, temperature and housing should be ensured.

## Written Work

1 Amphibians were the ancestors of reptiles.
2 Modern reptiles are land animals with scaly skins.
3 Their eggs are always laid on land.
4 Crocodiles feed on animals caught in the water.
5 The shell of a turtle protects its body.
6 Most lizards have four legs.

**LIVING THINGS HAVE YOUNG**

## THE PLANTS

### Demonstration Material

1   Any specimens to show vegetative reproduction: new plants growing from part of the parent, *e.g.* young bulbs, twigs kept in water which have developed adventitious roots, a potato with new shoots.

2   Spore cases, *e.g.* fern leaf or leaflet with spore cases, clubmoss spore cases, horsetail spore case, mushroom, or other toadstool such as puff ball or tree fungus, common moss with spore cases.

3   Seeds:
   *a* pine cone containing winged seeds
   *b* fruits and seeds, including any left over from the autumn collection, *e.g.* conkers, acorns, beans.

### Sample Link Questions

1   What are the two kinds of living things? (*Animals and plants*)
2   What are the three main ways in which animals can have young? (*By part of the parent growing into a new animal; by means of eggs; by means of tiny babies*)
3   How do *most* kinds of animal have young? (*By means of eggs*)
4   Which is the best way in which animals can have young? (*By means of tiny babies*)
5   What do we call plants which have no roots, stems or leaves? (*Simple plants*)
6   What are the three kinds of plants on land with roots, stems and leaves? (*Herbs, trees, shrubs*)
7   Do any simple plants have flowers, fruits and seeds? (*No*)
8   What do we call the part on a stem from which a new plant part grows? (*A bud*)
9   What else may grow from a bud apart from a branch stem with new leaves? (*Flowers, or a branch stem with both leaves and flowers*)
10   What do we find inside fruits? (*Seeds*)

**Relevant Information**

The main purpose of this lesson is to show that in the plant kingdom, there are three main ways of having young:

1 by the development of a new individual plant from a section of the parent;
2 by means of spores;
3 by means of seeds – which differ from spores in that they contain food reserves.

### 1 The Development of a New Individual from some Section of the Parent

In this process (known as vegetative reproduction) part of the parent plant develops as a new individual and may eventually become detached. There are several methods by which this kind of reproduction may occur.

A *Simple Plants* – plants without true roots, stems and leaves
a A one-celled plant simply divides, one individual becoming two – as with the very simplest of animals.
b Thread-like algae and fungi composed of strings of cells may increase by a few cells breaking off and developing into fresh filaments. Mushroom spawn, for example, is a compact mass of fungus threads (the mycelium), which are the body of the plant proper. This spawn, when broken up and given favourable conditions, develops new threads from the existing ones. It is from these threads that the spore cases grow. The spore cases, of course, constitute the edible part of the plant.
c Among the more highly evolved of the simple plants such as the seaweeds, new individuals may be developed from 'buds' formed on the parent plant. Mosses and liverworts may be grown from separate 'buds' on the old plant, or from pieces which have become detached.

B *Plants with Roots, Stems and Leaves* – herbs, trees and shrubs

Among the higher forms of plant life, vegetative propagation may take place in a variety of ways, *e.g.* by means of 'running' stems, by budding, from cuttings, and by grafting. In the majority of cases the new individual develops from part of the stem of the parent plant. There are plants, however, where reproduction can take place from a leaf or a root. Some of the ferns may reproduce by means of small buds developed on the leaves. The Indian fern, which is a decorative plant used by keepers of tropical aquaria, is an example of a fern which frequently reproduces from the leaves in this way. It is because the dandelion can reproduce from part of the severed root left in the ground that it is so difficult to eradicate from the garden.

## 2 Reproduction by means of Spores

### A Simple Plants

Spores may have one cell or several, and new plants develop from them either directly or indirectly. They are usually microscopic and produced in enormous numbers. Some of the giant puff balls produce quantities which have been estimated at about one hundred million million. Spores are of two kinds.

a *Those produced asexually* – without the fusion of male and female cells, but by the rapid division of cells. Simple asexual reproduction by means of spores is a step higher than vegetative reproduction, which is the only means available to the simplest of the simple plants. Asexual spores are reproduced by quite a number of the algae and fungi.

b *Those produced by sexual means* – as a result of the union of male and female cells. This type of reproduction is a step higher than asexual reproduction and requires in the plant the presence of male and female parts. The highest forms of algae and fungi are capable of it.

On some plants spores are formed on the body of the plant itself, and on others they are raised from the plant in some way. The brown patches seen on dry rot are the spore patches – sometimes misnamed fruiting bodies – and the brown dust found nearby is composed of scattered spores. The toadstools and mushrooms that we see above the ground are only the spore-bearing cases of fungi and do not constitute the main bodies of these plants. The main plants themselves consist of many tiny thread-like filaments feeding unseen below the surface.

Mosses and liverworts are higher forms of plants than algae and fungi, and, on these, sexually reproduced spores are raised in little capsules.

213

## B *Plants with True Roots, Stems and Leaves*

Among the herbs, trees and shrubs it is only in the lowest classes that reproduction by means of spores takes place. This applies to the following three classes of plants, present-day examples of which are herbs.

*a* Clubmosses and their relatives, which have spores in cone-shaped cases.

*b* Horsetails, which also have spores in cone-shaped cases.

*c* Ferns, which develop spores in cases on the undersides of their leaves.

The clubmosses, horsetails and ferns were the dominant plants at the time when the amphibians were the dominant animals. It was from those clubmosses, horsetails and ferns that coal beds originated.

Like seeds, spores may be dropped from plants or they may be expelled, *e.g.* puff ball. Again like seeds, they are distributed by the usual three agents, *i.e.* moving animals, moving air or moving water.

## 3 *Reproduction by means of Seeds*

Seeds are produced by sexual means and have one or more protective coats. They are many-celled structures and contain food reserves such as starch and oil. A spore does not have the same reserves of food as a seed and is unable to withstand conditions of adversity as a seed can. None of the simple plants have seeds. Seeds can be produced only by the higher classes of herbs, trees and shrubs. They develop in two ways.

*a By means of cones and seeds.* This was the most advanced method in the days of the giant reptiles. Present-day survivors of this class are certain trees and shrubs, and include the pines, firs, spruces, larches, redwoods, yews and junipers. Cones are not fruits. They are mostly woody, although on yews and junipers they are soft, fleshy and colourful.

*b By means of flowers, fruits and seeds.* This method is more highly evolved than that of the coniferous trees and shrubs and is the method of most of the herbs, trees and shrubs of today. The seeds are developed inside a fruit and are completely enclosed by the fruit wall.

The conditions under which spores and seeds will germinate vary according to species. Moisture and temperature are important factors. The seeds of aquatic herbs germinate when completely immersed in water, but those of land plants can drown owing to the insufficiency of the oxygen supply. Quite a number of seeds have a seed coat which keeps out water, and these will not germinate until the coat breaks or decays.

The illustrations for this lesson in the Pupils' Book show:

## 1 Reproduction from Part of the Parent

*Yeast*. This is a simple plant consisting of single cells or short chains of cells. Reproduction is by division from a single cell.

*Strawberry*. This is a herb which, at the end of the flowering season, produces branch stems (called *runners*). These stems elongate until they bend and touch the ground. When a node comes into contact with the ground, adventitious roots are formed, and these give rise to new plants.

*Willow*. This represents a family containing many trees and shrubs including the crack willow, cricket-bat willow, osier willow, weeping willow and others. The one illustrated is the goat willow which annually suffers severe and drastic pruning as a result of its flowers appearing in attractive clusters known as catkins. Adventitious roots frequently form on severed stems when placed in water.

## 2 Reproduction by means of Spores

*Puff Ball*. This simple plant is a common fungus whose spores are contained in the familiar round case. When ripe, the case bursts to release anything from 1000 to 2000 million spores. The plant itself feeds on dead materials.

*Moss*. This is one of the highest of the simple plants. The spores are raised in a stalked capsule which can easily be seen. When it is ripe, a tiny lid on the capsule opens and allows the spores to be shaken out and carried by the wind.

215

*Fern*. This represents the herbs which have young by means of spores. The cases containing the spores develop on the underside of the fern leaflets. The cases open when ripe, producing hundreds of millions of spores per plant.

## 3  Reproduction by means of Seeds

*Pine*. This is one of the most important of the conifers. The cone scales open when ripe, exposing the seeds which are equipped with wings to let the wind carry them more easily.

*Bean*. This represents herbs which have young by means of flowers, fruits and seeds. The fruit itself is the pod. If the two halves of a bean seed are parted, a small curved shoot with tiny leaves is exposed – a clear illustration of a seed consisting of a baby plant and a store of food.

*Oak*. This is a tree which has young by means of flowers, fruits and seeds. The flowers grow in clusters known as catkins. The fruit is the familiar acorn, and the seed is inside this. The cup which holds the acorn is formed from the overlapping scales from the base of the female flower. The fruit is green at first. Oaks are not regular in their habits, and seasons when acorns are scarce are not unusual.

## CODE

*Reproduction from part of the parent (General notes on Observations and Experiments)*

1  Examine any plant where this kind of reproduction has obviously taken place.
2  Observe any young bulbs growing from side of the parent bulb, *e.g.* daffodil. (Remove dead outer leaves if necessary.)
3  A potato is a swollen stem part. Place one on a jar of water, but not touching the water. Observe the development of branch stems and, from these, leaves and roots. Observe also the eventual development of small potatoes.
4  Observe adventitious roots which may have developed from the twigs of trees and shrubs kept in water, *e.g.* privet, willow, and rose.

5   Observe how sections from the stems of several freshwater herbs may develop fresh growth if left floating in a jar of water. Canadian pondweed (*Elodea Canadensis*) is a good example.

*Reproduction by means of spores. (General notes on Observations and Experiments)*

1   Observe any spore cases found on simple plants, *e.g.* fungus and alga, mosses and liverworts, and on herbs, *e.g.* clubmosses, horsetails and ferns.

2   Experiment to obtain spores from a fungus, *e.g.* squeeze a puffball fungus over a sheet of white paper, or remove the 'stalk' from a mushroom, and leave spore side down under a jar, on a sheet of white paper, to obtain a possible spore print.

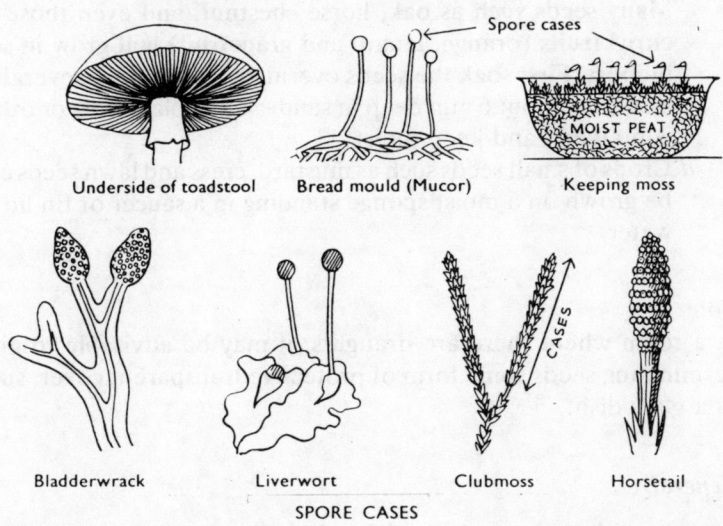

Underside of toadstool    Bread mould (Mucor)    Keeping moss

Bladderwrack    Liverwort    Clubmoss    Horsetail

SPORE CASES

Fungus parts should obviously not be left in the care of any child who is liable to taste them.

3   Observe how fungus plants such as moulds, rusts and smuts appear on living or dead things, and on foods with animal or plant content such as bread. Such fungus plants appear as a result of air-borne spores.

217

*Reproduction by means of seeds. (General notes on Observations and Experiments)*

1 Observe *a* the seeds exposed by an open pine cone
         *b* the seeds enclosed in various fruits.
2 Split open a bean seed and examine the baby plant inside.
3 Plant various seeds to grow.
   *a* Where it is desired to observe the formation of roots, stem and leaves from seeds such as pea and bean, soak the seeds overnight. Then insert them between moist blotting paper or absorbent tissue and the sides of a jar. It is important to ensure that there is always sufficient water in the jar, both for germination and for continuation of growth. Peat or even sawdust inside the blotting paper helps to hold it firm and will also store a reserve of water. (Also see CODE 2*c*, Lesson 27.)
   *b* Bought seeds may be grown according to the instructions on the packet.
   *c* Many seeds such as oak, horse chestnut, and even those of citrus fruits (orange, lemon and grapefruit) will grow in soil indoors. First soak the seeds overnight. Then plant several of one kind about 6 mm deep in sandy soil in plant pots or other containers, and keep moist
   *d* Crops of small seeds such as mustard, cress and lawn seeds can be grown on a moist sponge standing in a saucer or tin lid of water.

*Note*

In a room where there are draughts, it may be advisable to give germinating seeds some form of protective transparent cover, such as a glass dish.

*General*

Preserve selected specimens showing the development of new plants in 5% formaldehyde. (See Keeping Specimens for Observation Purposes, page 27 for use of copper acetate.)

**Written Work**

1 Some young plants may grow from part of the parent.
2 Many simple plants have young by means of spores.

3 A few herbs have spores, but most have <u>seeds</u>.
4 All <u>trees</u> and <u>shrubs</u> may have young by means of seeds.
5 Most seeds grow inside a <u>fruit</u>.
6 Some trees and shrubs have their seeds in <u>cones</u>.

## QUESTIONS ON LESSONS 13 TO 24

1 What are the three forms in which we find never-alive things? *Solid, liquid and gas*

2 Is coal gas just one gas or a mixture of gases? *A mixture*

3 In order to make a paste, you would mix a liquid with *either* a solid or a gas. Which? *A solid*

4 What do we call the mixture in which tiny bubbles of air are mixed with water? *Foam*

5 What is the liquid in which most things will dissolve? *Water*

6 What name do we give to a mixture in which something is dissolved? *A solution*

7 When something dissolves, it breaks up into the tiniest possible bits. What do we call these bits? *Molecules*

8 What is the commonest solid dissolved in sea water? *Salt (common)*

9 What carries the salt down to the sea? *Streams and rivers*

10 What forces streams and rivers to flow down towards the sea? *Gravity*

11 When salt water dries up, what is left behind? *Salt*

12 When a liquid freezes, does it become a solid or a gas? *A solid*

13 Is water vapour solid, liquid or gas? *Gas*

14 When water vapour freezes and falls from the air, what do we call it? *Snow*

15 When a solid melts, does it become a liquid or a gas? *A liquid*

16 When sugar is mixed in a cup of tea, does it melt or dissolve? *It dissolves*

17  Which liquid do all living things need? — *Water*

18  Some animals need more water than they find in their foods. How do they take in extra water? — *By drinking it*

19  What do we make to collect water from spaces in underground rocks? — *A well*

20  What do we make to store the water that we obtain from streams? — *A reservoir*

21  What do we call the wall that we build across a valley to make a reservoir? — *A dam*

22  What do we call the two points in a magnet at which its pull is greatest? — *Poles*

23  In a magnet are the poles near the ends or near the middle of the magnet? — *Near the ends*

24  Which two solids make good magnets? — *Iron and steel*

25  What are the four main needs of living things? — *Oxygen, food, to grow, to have young*

26  Which gas is used when we breathe, and when most things burn? — *Oxygen*

27  What do some animals have to help to take oxygen from the air? — *Lungs*

28  What do some animals have to help to take oxygen from the water? — *Gills*

29  Does a bud usually grow from a root, a stem or a leaf? — *A stem*

30  What may kill some buds on trees and shrubs during winter? — *Frost*

31  How do most kinds of animals have young? — *By eggs*

32  Which class of animals always cares for its young? — *The mammal class*

33  Which animals were the ancestors of the first reptiles? — *Amphibians*

34  Which animals were the ancestors of the first amphibians? — *Fish*

35  Where are the eggs of most amphibians laid – in water or on land? — *In water*

36  Where are the eggs of reptiles always laid? — *On land*

37  Which have four legs and a tail when they are adults – salamanders and newts, the frogs and toads, or the caecilians? — *Salamanders and newts*

38 The skins of some amphibians are smooth and moist. How do the skins of other amphibians feel when you touch them? *Rough and dry*

39 Which has a scaly skin – a lizard or a newt? *A lizard*

40 Of the simple plants, herbs, trees and shrubs, which are the ones that can have young by means of seeds? *The herbs, trees and shrubs*

# 25 HERBS, TREES AND SHRUBS WITH FLOWERS

## FLOWERS ON THEIR OWN AND FLOWERS IN CLUSTERS

### Demonstration Material

1 Any stem from which a solitary flower is growing, *e.g.* daffodil, tulip, wood anemone.

2 Any stem from which a cluster of flowers is growing, *e.g.* bluebell, horse chestnut, currant, cherry, almond, apple, lilac, laburnum, tree catkins, dandelion head.

### Sample Link Questions

1 What are the three main ways in which plants have young? (*By young plants growing from part of the parent; by means of spores; by means of seeds*)

2 Which is the best way for plants to have young? (*By means of seeds*)

3 The seeds of herbs and the seeds of most trees and shrubs always grow inside something. What? (*A fruit*)

4 From which special part on a plant does a fruit grow? (*A flower*)

5 From what part on a stem does a flower grow? (*A bud*)

6 Where are buds usually found – on roots, stems or leaves? (*On stems*)

7 Whereabouts on a stem do buds grow? (*At the end and on the sides*)

8 At which points on a stem do we find most of the side buds growing? (*Where a leaf joins the stem*)

9 What kinds of plant parts grow from buds? (*Sections of stem with new leaves; flowers; sections of stem with both leaves and flowers*)

10 Name the three main kinds of plants with roots, stems and leaves. (*Herbs, trees, shrubs*)

### Relevant Information

The main points covered by this lesson are:

1 Flowers can only be grown by plants with roots, stems and leaves. This means that they can be grown on most, but not all, herbs, trees and shrubs.

2 Flowers are arranged on a stem in two main ways.

   *a* On some plants each flower grows separately some distance from the others.

   *b* On other plants they grow together in clusters.

Different kinds of clusters are dealt with in a lesson in Book 3. The main flower parts are dealt with in Book 4.

A flower is the reproductive structure of those plants which can have young by means of flowers, fruits and seeds. Plants which can reproduce in this way exclude all simple plants but include most of the present-day herbs, trees and shrubs.

Present-day herbs, trees and shrubs, *i.e.* those plants which have true roots, stems and leaves, are classed according to the method by which they reproduce themselves.

1 *Herbs which can have young by means of spores* – the club mosses, horsetails and ferns. Clubmosses, horsetails and ferns were dominant plants during the time when amphibians were the dominant animals.

2 *Trees and shrubs which can have young by means of cones and seeds* – gymnosperms. Reproduction by means of seeds is more advanced than reproduction by spores. The cone-bearers were the dominant plants during the time when reptiles were the dominant animals.

3 *Herbs, trees and shrubs which can have young by means of flowers, fruits and seeds* – angiosperms. Reproduction by means of flowers, fruits and seeds is more advanced than reproduction

222

by means of cones and seeds. The flowering herbs, trees and shrubs have become the dominant plants in the world, just as mammals and birds have become the dominant animals.

Not all angiosperms produce flowers every year. Biennial herbs, for example, build up a food store during the first year and produce flowers during the second.

Trees and shrubs may live many years before they bear flowers. The birch and the sweet chestnut, for example, may be about twenty-five years old before they are sufficiently mature to do so. The ash may be between forty and fifty, and the oak may be sixty or seventy years old before it begins to reproduce.

Some plants flower infrequently – for example, the cacti, some of which may produce flowers only about every twelve years or so. Then again, some plants manage to survive in conditions which are not entirely suitable to them and in consequence are unable to produce flowers at all. The flowers of duckweeds, for example, are seldom seen in Britain. Duckweeds are reputed to be the smallest plants which can have flowers, fruits and seeds.

Although spring and summer are generally accepted as the seasons for flowers, there are flowers at all seasons. Different heather shrubs, for example, produce their flowers at different times during the year.

Flowers are believed to have evolved from cones, and it is for this reason that the cones of coniferous trees and shrubs are sometimes referred to as flowers.

Flowers are looked upon as being specialised sections of stem. They develop from buds in the following ways.

1  *From a flower bud* either a single flower or a cluster of flower buds will develop, according to the habit of the plant.
2  *From a mixed bud* a section of stem bearing both leaves and flowers will develop.


*Arrangement of Flowers on the Stem*
(Generally termed *inflorescence*)

A wealth of information has been presented by botanists as a result of their studies of flowers, inflorescence, and flower structure, and a large number of terms coined. The following is a simplified explanation of the different ways in which flowers are arranged on a stem, together with the botanical terms used.

### FLOWERS IN CLUSTERS
#### WITH STALKS

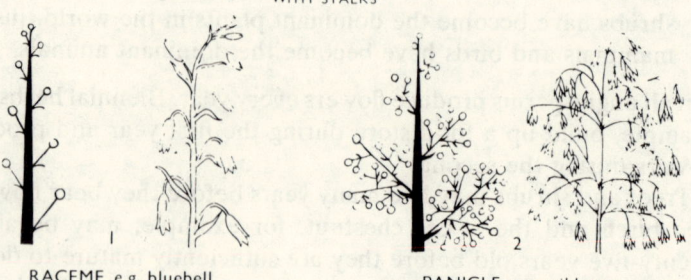

RACEME e.g. bluebell     1     PANICLE e.g. wild oat     2

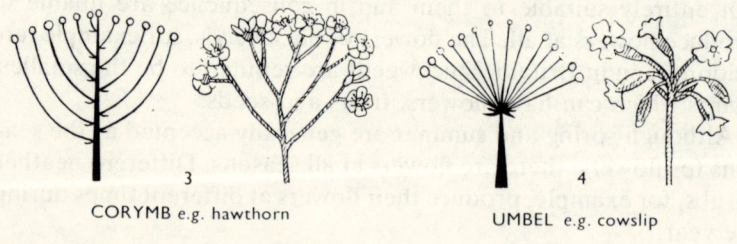

CORYMB e.g. hawthorn     3     UMBEL e.g. cowslip     4

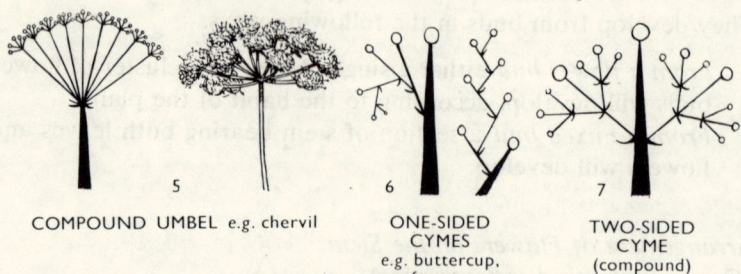

COMPOUND UMBEL e.g. chervil     5

ONE-SIDED CYMES
e.g. buttercup,
water forget-me-not     6

TWO-SIDED CYME
(compound)
e.g. chickweed     7

The first five examples are developments of the raceme type.
The cymes (6 and 7) are side-branching types of inflorescence.

Simple floral diagrams are on the left in examples 1-5. The thick lines show the main flower stem. Flowers may be opposite or alternate on branch stems.
One-sided or simple cymes often grow on plants with alternate leaves. Two-sided or compound cymes often grow on those with opposite leaves.
The reduced leaves shown in the diagrams are called *bracts*.

## Flowers on their Own

On some flower-bearing plants the flowers are solitary, being carried alone on a stem which rises straight up from the ground. On others the flowers are borne singly in the axils of the leaves.

FLOWERS ON THEIR OWN

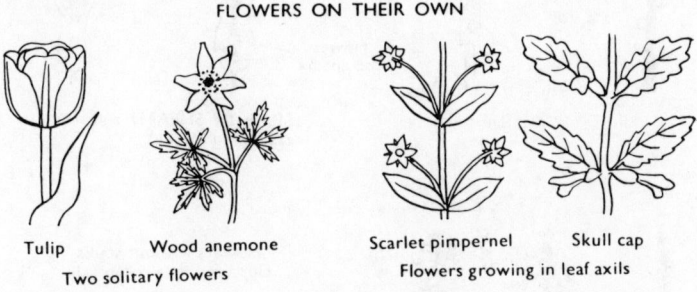

Tulip            Wood anemone              Scarlet pimpernel      Skull cap
Two solitary flowers                    Flowers growing in leaf axils

## Flowers in Clusters

On some flower-bearing plants several or many flowers grow together in a cluster.

On some clusters each individual flower grows from the stem on a short branch stem or stalk. Racemes are typical formations on which this occurs.

On other clusters the flowers are without a stalk (a condition known as *sessile*), and grow directly from the stem. Spikes are typical of this kind of formation. (Illustrated overleaf.)

A catkin is sometimes referred to as a spike. A catkin differs, however, in that all its flowers are of one and the same sex – either all the flowers are male, possessing stamens but no pistils, or they are all female, possessing pistils but no stamens.

*Composite Heads of Flowers*. This type of inflorescence is known as a capitulum. It consists of a compact cluster of flowers crowded on to a flat or conical receptacle. These stalkless flowers are known as florets, and there are two kinds. Some are tubular, and some are strap-shaped. On the dandelion they are all strap-shaped. On the burdock they are all tube-shaped. On daisies there are both kinds – strap-shaped (ray florets) on the outside, and tube-shaped (disc florets) on the inside. On a single daisy flower head, there may be about 250 flowers.

The composite family (*compositae*) number more than 13 000

225

### FLOWERS IN CLUSTERS
#### WITHOUT STALKS

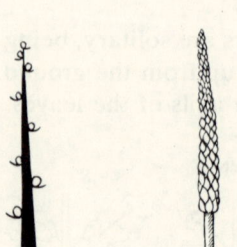

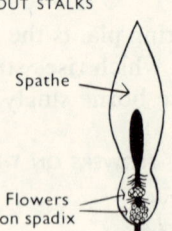

Spathe

Flowers on spadix

SPIKE e.g. plantain

SPIKE IN SHEATH e.g. arum
(spadix in spathe)

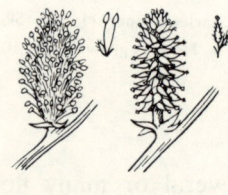

Flowers without stalks clustered on a receptacle

CATKIN e.g. willow

COMPOSITE FLOWER e.g. daisy

## FLOWERS WITHOUT STALKS CLUSTERED ON A RECEPTACLE

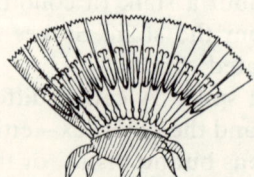

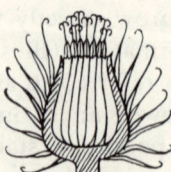

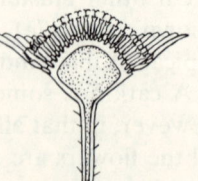

A section of the head of a dandelion showing a row of strap-shaped florets

Section through a burdock head showing tube-shaped florets

Section through a daisy head showing both strap- and tube-shaped florets

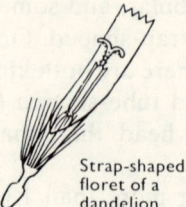

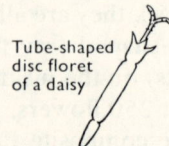

Strap-shaped ray floret of daisy

Tube-shaped disc floret of a daisy

Strap-shaped floret of a dandelion

226

species, and they are the most widely distributed of the flowering plants. In the northern hemisphere they are mostly herbs, but in countries such as South America some are shrubs. In Great Britain there are over 100 species, including such well-known plants as the aster, cornflower, groundsel, coltsfoot, fleabane, butterbur, yarrow, tansy, hawksweed, the thistles, and many others. Many of them develop plumed fruits to facilitate wind-dispersal.

Clusters of flowers have a certain advantage over solitary flowers, in that there are more of them; in a compact mass they are more conspicuous, and the chances of pollination and eventual reproduction of the plant are thereby increased. The flower spike of a palm tree may be 6 metres in length, with a cluster of some 200 000 flowers.

Normally, a stem is lengthened by new growth developing from the terminal bud, but once a terminal bud has produced a flower, no further lengthening of that particular stem takes place.

## CODE

1   Collect flowers which are brought into the classroom into two groups:
    *a* flowers which grow on their own
    *b* flowers which grow in clusters
2   Observe how, in some clusters, it is possible to see each flower emerging from its own individual bud in the cluster, *e.g.* apple, currant, horse chestnut, laburnum.
3   Observe the individual flowers in a flower cluster.
4   Observe as many different kinds of flowers as possible, including those which are sometimes overlooked as such, *e.g.* grass flowers, and flowers on common garden 'weeds' such as plantain and groundsel.
5   Collect specimens of small flowers for mounting in notebooks with cellulose tape. For a more permanent collection, press them first, and mount on stiff card, or – more effectively – on a sheet of glass or perspex. Covering completely with self-adhesive clear plastic will preserve them.

**Written Work**

1   Only <u>herbs</u>, <u>trees</u> and <u>shrubs</u> can have flowers.

2   Flowers never grow on simple plants.
3   A plant must have flowers before fruits and seeds.
4   On some plants each flower grows on its own.
5   On other plants the flowers grow in clusters.

# 26   LEAVES OF HERBS, TREES AND SHRUBS

## LEAVES AND LEAFLETS

### Demonstration Material

1   As many as possible of the following:
  a A section of a stem from which is growing a simple leaf without a stalk (a sessile leaf).
  b A section of a stem from which is growing a simple leaf with a stalk.
  c A section of a stem from which is growing a compound leaf with leaflets radiating from the end of the stalk.
  d A section of a stem from which is growing a compound leaf with leaflets growing from the side of the stalk.
2   Any stem on which an axillary bud can be clearly seen between the leaf stalk and the stem.

### Sample Link Questions

1   Which three kinds of plants can have flowers? (*Herbs, trees, shrubs*)
2   Which plants always keep their stems above ground during the winter? (*The trees and shrubs*)
3   In what way do the stems of trees and shrubs differ from the stems of herbs? (*The stems of trees and shrubs are woody*)
4   What do we call trees and shrubs which have green leaves in winter? (*Evergreen*)
5   What do we call trees and shrubs which have no green leaves in winter? (*Deciduous*)
6   What do we call leaves that grow from a stem one at a time? (*Alternate leaves*)

228

7   What do we call leaves that grow from a stem two at a time? (*Opposite leaves*)
8   From which part on a herb, tree or shrub do buds grow? (*The stem*)
9   Whereabouts on the side of a stem do most side buds grow? (*Where a leaf joins the stem*)
10  What is it that has leaves on it and that grows out of a bud? (*A stem*)

## Relevant Information

Leaves are an acknowledged aid to the recognition of different herbs, trees and shrubs. A lesson in Pupils' Book 1 dealt with the two main ways in which leaves may be arranged on the stems of deciduous trees, *i.e.* opposite and alternate leaves.

In this lesson the two main kinds of leaves are dealt with, *i.e.*

1   *Simple leaves*
    *a* without a leaf stalk
    *b* with a leaf stalk
2   *Compound leaves*
    *a* with leaflets radiating from the end of the stalk
    *b* with leaflets growing from the side of the stalk

The different edges and veins of leaves are dealt with in a lesson in Book 3.

Leaves have been described as the factories of a plant. In general their three main purposes are:

1   *To obtain oxygen.* Fresh air containing oxygen flows in through the open pores (stomata). The oxygen is absorbed by diffusion, and the remainder of the air (mainly nitrogen) is released, together with any excess carbon dioxide formed during respiration. The process is fundamentally the same as in the animal kingdom.
2   *To obtain food.* From the carbon dioxide in the air, the carbon is extracted and combined with water conducted from the roots, utilising the energy of light during the process. This manufacture of simple carbohydrates is known as *photosynthesis.* *Chlorophyll* – the green pigment present in most of the cells of a leaf – is necessary for photosynthesis.
3   *To permit transpiration.* Water taken in by the roots is con-

ducted to the leaves by means of the stem, carrying with it dissolved salts. Excess water evaporates through the open pores, leaving the salts behind.

Although these are the main functions of a leaf, certain leaves on certain plants may be modified for other purposes. Bud scales, for example, are modified leaves; some spines and some tendrils are modified leaves, *e.g.* tendrils on a pea plant. Certain leaves may be modified for storage of food, as they are in the bulbs, and the herbs with succulent leaves.

Leaves vary in shape and size, from the thin narrow leaves of the grasses to the almost circular leaves of the nasturtium, but they are generally arranged on the stem so as to obtain the most sunshine and air. The point on the stem to which the leaves are attached is called the *node*.

## Arrangement of Leaves on Stems

1  *Alternate*. In this arrangement, there is one leaf at each node. (This is the most common arrangement.)
2  *Opposite*. In this arrangement, there are two leaves at each node.
3  *Whorled*. In this arrangement, there is a group of three or more leaves at a node. On stems where the arrangement of the leaves is alternate, successive leaves usually form an ascending spiral round the stem which enables the bulk of the leaf surface to be presented to the light. On stems where the arrangement of the leaves is opposite, each pair of leaves is normally at right angles to the pair above and the pair below, thus covering the space between the pair above and the pair below, to enable the bulk of the leaf surface to be presented to the light. Sometimes the leaf

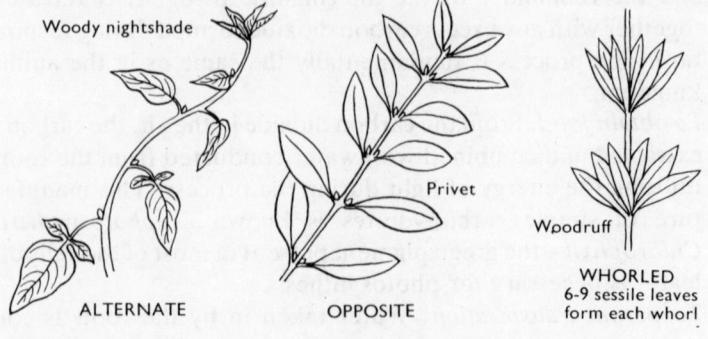

Woody nightshade

ALTERNATE

Privet

OPPOSITE

Woodruff

WHORLED
6-9 sessile leaves
form each whorl

230

stalks of the lower leaves lengthen so that the blades of these leaves are extended farther than those above. On those comparatively few stems where the arrangement is whorled the leaves are usually evenly spaced around the stem, again for the best possible presentation.

A typical leaf is a broad and thin outgrowth from the stem. The flattened portions of leaves are referred to as blades, and although on some plants each blade grows directly from the stem, in most cases it is attached to the stem by a short stalk (the *petiole*). On grass plants petioles are flattened into sheaths which clasp the stem. Leaves which have no petiole are referred to as *sessile*.

Blade

Lobes

Veins

Stalk (petiole)

Without stalk (sessile)    With stalk

## Simple Leaves

A simple leaf consists of one blade, the edge of which may be smooth, toothed, or lobed. It may grow directly from the stem or be attached to it by a stalk.

## Compound Leaves

Compound leaves have several blades growing from the same stalk. The blades are known as leaflets, and the different arrangements of these leaflets results in two kinds of compound leaves.

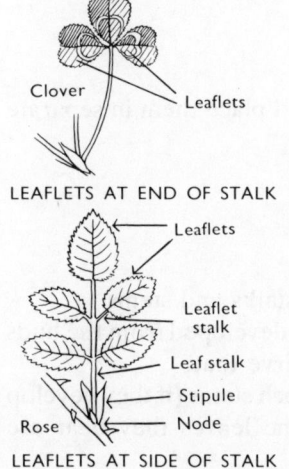

Clover

Leaflets

LEAFLETS AT END OF STALK

Leaflets

Leaflet stalk

Leaf stalk

Stipule

Rose    Node

LEAFLETS AT SIDE OF STALK

1 Leaves on which leaflets are attached to a common point at the end of the stalk, as in the clover and the horse chestnut. This type is known as palmately compound.

2 Leaves on which leaflets are attached to the sides of the stalk, as on the potato, rose and ash. This type is known as pinnately compound. On some plants the leaves may be more than once pinnately compound, so that each leaflet is itself divided into sub-leaflets (*e.g.* ferns).

231

As will be seen from a rose leaf, it is possible for the separate leaflets themselves to have stalks. This kind of compound arrangement gives rise to the problem of how to distinguish between a compound leaf on which each leaflet has its own stalk, and a branch stem with simple leaves. There are two easy ways of telling:

1 Axillary buds develop in the upper angle between a leaf stalk and the stem from which it grows. They do not develop in the angle between the stalk of a leaflet and the main leaf stalk.
2 Some plants have small green outgrowths where the leaf stalk joins the stem. These outgrowths from the stem are known as *stipules*. They may be seen clearly where the stalk of the compound rose leaf joins the stem. Stipules do not grow where the stalk of a leaflet joins the main stalk of the compound leaf. (Stipules may protect the developing buds. In plants such as garden peas, they help in the manufacture of food. Some plants have stipules which fall off as soon as the bud opens, and many plants lack them altogether.)

*Note*

Not every plant has all its leaves alike. On charlock, clustered bell-flower and curled dock, for example, the lower leaves are stalked, while the upper leaves are sessile. On the dahlia, potato and sheep's scabious, the lower leaves are simple and the upper leaves are compound.

**CODE**

1 Collect different stems with leaves and place them in separate jars of water, labelled:
   *a* simple leaves without a stalk
   *b* simple leaves with a stalk
   *c* compound leaves with end leaflets
   *d* compound leaves with side leaflets
2 Observe buds growing between leaf stalks and stems.
3 Observe any branch stems which have developed from the buds or twigs collected in winter, and observe that:
   *a* where buds were opposite, then branch stems (if they develop properly) are also opposite, and the leaves they bear are opposite too

232

MOUNTING AND PRESERVING LEAVES

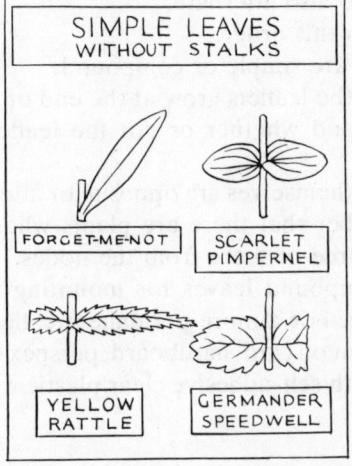

SIMPLE LEAVES
WITHOUT STALKS

FORGET-ME-NOT

SCARLET
PIMPERNEL

YELLOW
RATTLE

GERMANDER
SPEEDWELL

SIMPLE LEAVES
WITH STALKS

LOMBARDY
POPLAR

BEECH

LESSER
CELANDINE

WATER LILY

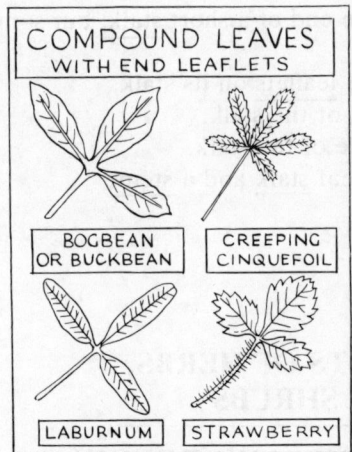

COMPOUND LEAVES
WITH END LEAFLETS

BOGBEAN
OR BUCKBEAN

CREEPING
CINQUEFOIL

LABURNUM

STRAWBERRY

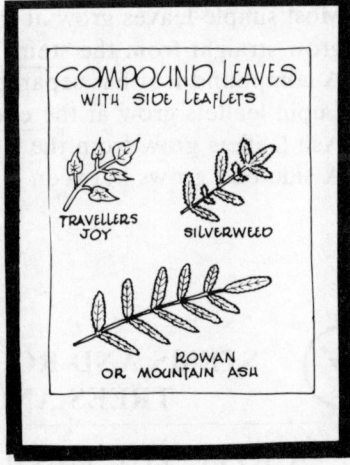

COMPOUND LEAVES
WITH SIDE LEAFLETS

TRAVELLERS
JOY

SILVERWEED

ROWAN
OR MOUNTAIN ASH

CASE FOR PRESERVING LEAVES

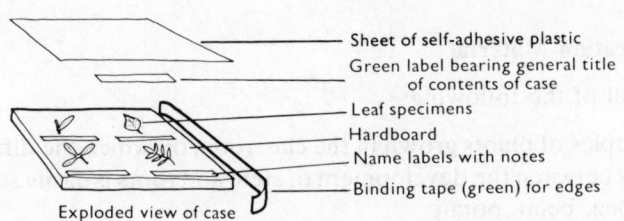

Sheet of self-adhesive plastic

Green label bearing general title
of contents of case

Leaf specimens

Hardboard

Name labels with notes

Binding tape (green) for edges

Exploded view of case

*b* where buds were alternate, then branch stems are alternate, and the leaves they bear are also alternate.

4 Observe a stem with leaves in this order:

*a* observe whether the leaves are simple or compound;

*b* if compound, note whether the leaflets grow at the end or at the sides of the leaf stalk and whether or not the leaflets themselves have stalks.

*c* Observe whether the leaves themselves are opposite or alternate on the stem. (Remember that there are plants whose leaves grow in clusters of three or more from the nodes.)

5 Collect small simple and compound leaves for mounting in notebooks with cellulose tape. For a more permanent collection, press them first, and mount on card, hardboard, perspex or glass. Covering completely with self-adhesive clear plastic will preserve them.

**Written Work**

1 Most simple leaves grow at the end of a short <u>stalk</u>, but some grow straight from the <u>stem</u>.

2 A compound leaf has separate <u>leaflets</u> on its stalk.

3 Lupin leaflets grow at the <u>end</u> of the stalk.

4 Ash leaflets grow from the <u>side</u> of the stalk.

5 A side bud grows between a leaf <u>stalk</u> and a <u>stem</u>.

 **27** **STEMS AND ROOTS OF HERBS, TREES AND SHRUBS**

## ABOVE THE GROUND AND BELOW

**Demonstration Material**

Any or all of the following:

1 Examples of plants grown in the classroom on which the differences between the development of stem and roots is easily seen, *e.g.* pea, bean, potato.

2   Specimens to illustrate any of the following:
    *a* Twining stems, *e.g.* hop, dodder, honeysuckle, woody night-
       shade, bindweed (great and small).
    *b* Stems with tendrils, *e.g.* pea (sweet and garden), Virginia
       creeper, white bryony, vetch (common, bush and meadow),
       passion flower, grape vine, garden nasturtium, clematis. In the
       last two cases the leaf stalk acts as a tendril.
    *c* Stems with hooks, *e.g.* blackberry, raspberry, loganberry,
       climbing and wild roses, goosegrass.
    *d* Prostrate or creeping stems, *e.g.* strawberry, ground ivy (dis-
       tinct from climbing ivy), creeping Jenny, lesser periwinkle,
       cinquefoil, silverweed, houseleek, iris, Solomon's seal. In the
       last two cases the stem is swollen for food storage.
3   Specimens to illustrate any of the following:
    *a* Underground 'running' stems, *e.g.* mint, lily of the valley,
       coltsfoot, wood sorrel, nettle, horsetail, marram grass, sand
       vetch, woodruff, couch grass, bracken.
    *b* Underground food storage stems (swollen stems), *e.g.* potato,
       artichoke, cuckoo pint, a corm, a rootstock.
4   A section of ivy stem showing adventitious roots.

## Sample Link Questions

1   What are the three kinds of plants with roots, stems and leaves?
    (*Herbs, trees and shrubs*)
2   Which two kinds of plants have woody stems? (*Trees and
    shrubs*)
3   What is the difference between a tree and a shrub? (*A tree has
    one main stem growing from the roots; a shrub has two or more
    main stems*)
4   From what part of a herb, tree or shrub do buds usually grow?
    (*The stem*)
5   What do we call leaves that grow from the stem one at a time?
    (*Alternate*)
6   What do we call leaves which grow from the stem two at a time?
    (*Opposite*)
7   What do we call a leaf which grows alone on a stalk? (*Simple*)
8   Which kinds of leaf has several leaflets on a stalk? (*Compound*)
9   What are the two kinds of compound leaves? (*Leaves on which
    the leaflets grow from the end of the stalk; leaves on which the
    leaflets grow from the side of the stalk*)

## 10 What part of a plant is a twig? (*A branch stem*)

### Relevant Information

The purpose of this lesson is to observe the variety of stems and roots which grow above the ground and below the ground.

Most of the plants normally observed by children have true roots, stems and leaves. These are the characteristics of the land plant, for with them, it can obtain the essential water from the soil and raise its leaves into sunlight and fresh air.

The functions of the land plant may be similar to those of the aquatic plant, but water and land offer very diverse conditions. To the true land plants – the herbs, trees and shrubs – the stem is an essential part, and roots and leaves may be looked upon as outgrowths from the stem, developed to serve specialised purposes.

### Stems

The stem of a herb, tree or shrub is a very important part and, together with any other appendages including leaves, is often referred to as the shoot. The main purposes of the stem are:

1    to conduct materials between roots and leaves
2    to raise the leaves into the sunlight and air, and to support them. In plants with flowers, stems carry the flowers, and support them in favourable positions for pollination.

The stems of trees and shrubs are woody and remain above ground level during winter, whereas the stems of herbs are comparatively thin and soft, and do not persist above the ground during winter. Stems branch on all trees and shrubs and on most of the herbs. Branch stems are referred to as branches or twigs.

Stems vary a great deal, both in their growth and in their form. The stem of a plant such as the dandelion or carrot is so short that the leaves appear to rise almost directly from the apex of the root.

Many stems need no support. Of those that are too weak to support themselves, some trail over the surface of the ground, *e.g.* creeping Jenny (moneywort). Others can climb and twine, and have developed various methods of doing this. In some, the stem twines round some kind of support – often another plant stem. On others, tendrils and hooks have been developed to help to keep the climbing stems supported.

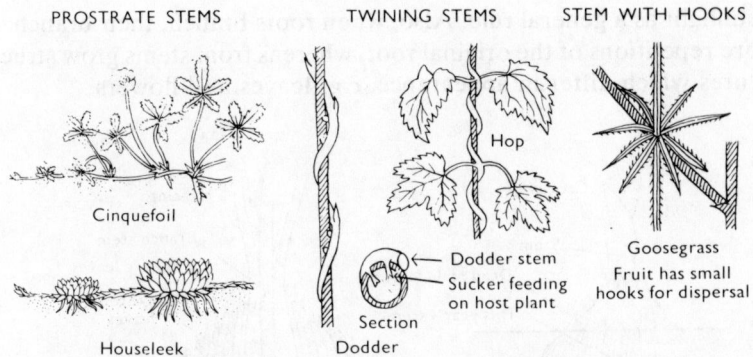

PROSTRATE STEMS — Cinquefoil — Houseleek

TWINING STEMS — Hop — Dodder stem — Sucker feeding on host plant — Section — Dodder

STEM WITH HOOKS — Goosegrass — Fruit has small hooks for dispersal

Hooks are outgrowths from the stem, but tendrils may be modified parts. The pea plant tendrils are modified leaves; the tendrils of the grape are modified stems, and the tendrils of the garden nasturtium and clematis are modified leaf stalks.

Twining stems do not all twine in the same direction. Bindweed, for example, twines anti-clockwise, while the honeysuckle twines in a clockwise direction.

In plants which live for more than a year (biennials and perennials) stems may store food. In most perennial herbs, the main stem has become an underground one, sending up leaves, or branch stems bearing leaves, each year. This underground stem is often confusingly termed 'rootstock'. A rhizome is a horizontal underground stem. A bulb is a stem modified for food storage, and consists of a very short compact stem with a cluster of thick fleshy leaves enclosing a terminal bud. In a corm all the food is deposited in the stem itself, and there are only a few scaly leaves on the surface of the corm. The stems of the iris and Solomon's seal, which are rhizomes storing food, are not completely underground. They grow along the ground, partly covered by soil, so that the leaves can grow above ground level. Underground stems can usually be distinguished from roots by the buds which grow from them. The growing point of a stem is also surrounded by buds or young leaves, whereas the growing point of a root is covered with a cap. Stems may grow along the top of the ground, or under the ground, or up towards the sunlight. They do not grow down away from the sun.

## Roots

Roots differ from stems in that they grow downwards away from the

237

sunlight as a general rule. Also, when roots branch, their branches are repetitions of the original root, whereas from stems grow structures which differ in appearance, *e.g.* leaves and flowers.

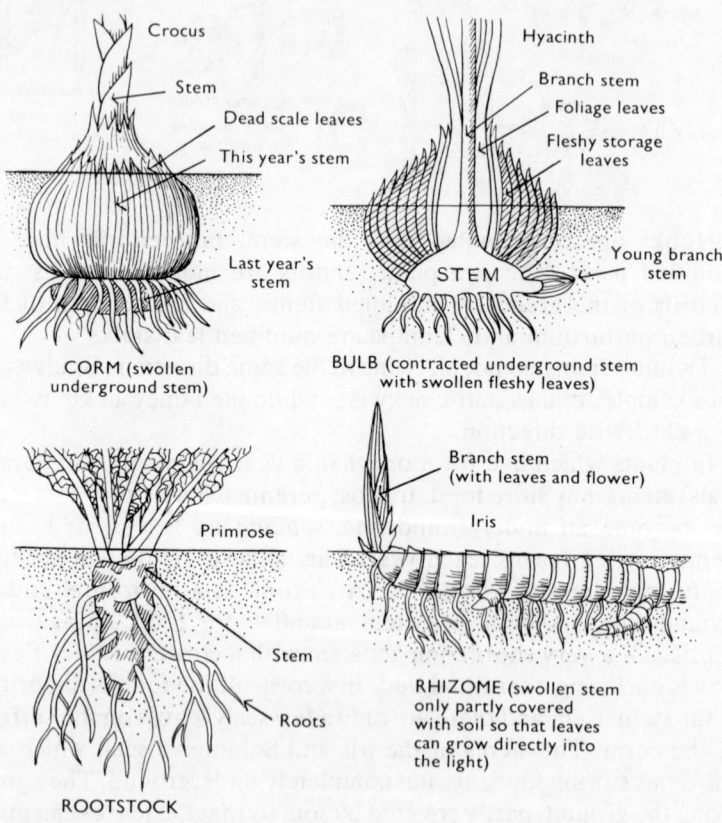

The main purpose of roots is to take in water and other foods. These consist chiefly of salts, which are dissolved in the water. The solution is absorbed into the root hairs by means of *osmosis*.

The second purpose of roots is to hold the plant in position – usually in soil, where the plant is then safeguarded against removal by wind. The roots of parasitic plants, such as mistletoe and dodder, enter the stem of the host plant. On floating aquatic herbs, such as water soldier, frogbit and duckweed, the roots do not hold the plant in any fixed position.

The two main root systems are:

1  *Fibrous*. Most roots are fibrous or bushy, spreading out below the surface of the soil, *e.g.* grass roots.
2  *Tap roots*. These are less common than fibrous roots, and reach deeply into the soil. Their branch roots are generally weaker and fewer than those found in a fibrous root system.

Many herbs, trees and shrubs have a root system which is intermediate between these two.

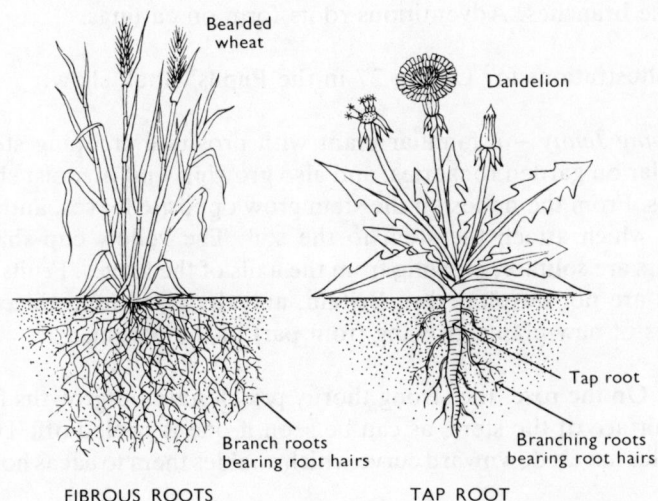

Bearded wheat

Dandelion

Tap root

Branch roots bearing root hairs

Branching roots bearing root hairs

FIBROUS ROOTS

TAP ROOT

On some herbs part of the root serves as a food store. Where the root system is fibrous, tubers may be developed, *e.g.* dahlia, and lesser celandine. On taproot systems, swollen organs may develop as on the carrot, swede, turnip, beetroot and parsnip.

Not all roots are below soil level. Many tropical orchids, for example, would have insufficient light if they grew on the ground. Some orchids are therefore adapted to growing from seeds which germinate in soil accumulated in the cracks of tree bark. Following germination, the roots are aerial, and take the moisture which they need from the humid air about them. Dodders are parasitic plants which germinate in the normal way. Their stems, which vary in colour from yellow to red, climb by twining round the stems of various other plants. From the climbing stem, 'sucker' roots enter the stem of the host upon which the dodder then becomes parasitic.

239

Leaves, being unnecessary, do not develop. When the plant is mature, the ground roots die off. Mistletoe on the other hand is a shrub whose roots are adapted for penetration of the stem of the tree upon which it becomes semi-parasitic. It does not have ground roots and retains the use of its leaves.

Adventitious roots are those which appear on the leaves of a plant or from some unusual part of the stem. The aerial roots of ivy, and those which form on a strawberry runner are adventitious. Banyan trees in India grow roots from their horizontal branches which grow down to the ground, become thickened, and thus serve as supports for the branches. Adventitious roots form on cuttings.

The illustrations for Lesson 27 in the Pupils' Book show:

*Creeping Jenny* – a familiar plant with prostrate creeping stems, popular on garden rockeries, and also growing wild in moist shady places. From the nodes on the stem grow opposite leaves, and also roots which attach the stem to the soil. The yellow cup-shaped flowers are solitary, growing from the axils of the leaves. Fruits and seeds are not developed in Britain, and the plant reproduces by means of new plants growing from part of the parent.

*Rose*. On the rose, the strong thorny prickles are outgrowths from the surface of the stem, as can be seen if one is peeled off. These prickles have a downward curve which enables them to act as hooks.

*Pea*. These plants, which are mainly annual herbs, climb by means of tendrils which are modified leaves.

*Great Bindweed*. This herb, with its stems twining in an anti-clockwise direction amongst the hedgerow plants, is a member of the convolvulus family. The large trumpet-shaped flowers are usually white, although pink ones do occur.

*Ivy* – an evergreen shrub with weak climbing stems held in place by adventitious roots growing from the sides of the stems. These roots also take in moisture – chiefly from the air itself.

*The Vines*. These include grape vines and Virginia creepers. The former climb by means of branch stems modified into tendrils. On the latter the ends of the tendril become adhesive with mucilage when in contact with a rough surface. These 'sticky fingers' enable the plant to climb walls. The virginia creeper illustrated is called *Ampelopsis*. It is a deciduous shrub with simple leaves. There is another very common species with compound leaves divided into leaflets at the end of the stalk.

*Potato*. The potato plant produces three types of stems:

*a* the aerial stem, which is the normal growth above the ground
*b* the rhizomes (surface or underground stems), which arise from the axils between the stem and the simple leaves which are produced by the plant near to the ground
*c* the edible stem tuber. This is really the end of the rhizome which swells in due season for the purpose of food storage.

If the ground is soft enough, the rhizomes become underground stems, but where the ground is hard, they will lie on the surface and produce small greenish potatoes. Earthing up by agriculturists encourages the underground growth of the rhizomes.

The 'eyes' on a potato are reduced buds in the axils of tiny scale-like leaves. It is from these buds that the aerial stems develop when a potato is planted – a good example of the growth of a new plant from part of the parent. Horticulturists plant small potatoes, and these are often termed 'seed potatoes', but the term is obviously a misnomer.

Although on the lower part of the aerial stem near to the ground there are simple leaves, the upper part of the aerial stem bears large compound leaves.

The potato is a stem tuber and should not be confused with the sweet potato which is a root tuber.

*Mint*. The mint family consists of perennial herbs with creeping underground stems, from which grow upright branch stems. The family includes the well-known garden varieties as well as several wild mints of which water mint is the commonest. The leaves are usually strongly aromatic and are a source of valuable oils. Flowers in most cases are lilac in colour and appear in clusters.

## CODE

1    Observe general differences between roots and stems, *e.g.* buds and leaves grow from stems; branch roots and root hairs grow from roots.

2   Demonstrate the general tendency for roots to grow downwards away from the light, and for stems to grow towards the light.

   *a* The leaves of herbs grown in pots on a window sill tend to turn towards the light. Turn pot so that the leaves no longer face the light, and observe how eventually they are once more turned towards the light.

   *b* Grow a plant in darkness, leaving a source of light for the stem to grow towards. A potato grown on a pot of water in a closed shoe-box will illustrate this. If a hole is cut in one end of the box, and partitions fixed to make things a little more difficult, the deliberate progress of the stem towards the light can be observed.

A STEM GROWING TOWARDS THE SOURCE OF LIGHT

Potato on a jar of water

Hole cut to allow light to enter

Partitions

(The box should, of course, be covered by a lid)

   *c* Grow bean or pea seeds exposed so that it is possible to observe that no matter which way up the seeds are planted, the roots tend to grow downwards and the stems upwards.

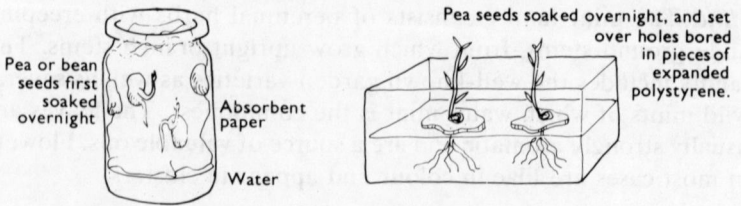

Pea or bean seeds first soaked overnight

Absorbent paper

Water

Pea seeds soaked overnight, and set over holes bored in pieces of expanded polystyrene.

GROWING PEA OR BEAN SEEDS EXPOSED

The absorbent paper must be moist enough to ensure germination and continuance of growth. Peat, or even sawdust inside the paper will store a reserve of water and hold the paper in position.

3 Observe twining stems and stems with tendrils.

4 Observe that hooks on bramble not only help the stem to climb but protect it from being eaten.

5 Observe underground 'runners' and prostrate stems which lie along the top of the ground.

6 Preserve stem sections showing tendrils or hooks in specimen tubes or screw-topped jars containing 5% formaldehyde. (See 'Keeping Specimens for Observation Purposes' page 27 for use of copper acetate.)

*Note*

Suitable herbs with climbing stems, *e.g.* pea and garden nasturtium, may be grown in pots of soil on a window sill with a twig or cane to act as a support for the climbing stems.

**Written Work**

1 Most stems grow toward the light.

2 Most roots grow away from the light.

3 Some weak stems grow along the ground.

4 Some climbing stems have hooks or tendrils.

5 Underground stems send up branch stems.

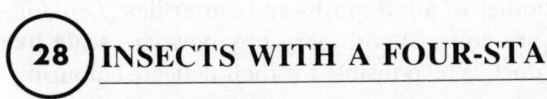

# 28 INSECTS WITH A FOUR-STAGE LIFE

## MOTHS AND BUTTERFLIES

**Demonstration Material**

1 Any living or preserved examples of:
   *a* moth or butterfly eggs
   *b* moth or butterfly larvae (caterpillars)
   *c* moth or butterfly pupae (chrysalises or chrysalides).

2 Any living or mounted example of:
   *a* an adult moth
   *b* an adult butterfly.

## Sample Link Questions

1 What name do we give to animals which always have six legs when they are adults? (*Insects*)
2 What are the stages in the life of an insect with a three-stage life? (*Egg, larva, adult*)
3 What are the stages in the life of an insect with a four-stage life? (*Egg, larva, pupa, adult*)
4 Which is the only stage of life in which an insect grows bigger? (*The larval stage*)
5 Which insect looks like the adult when it is a larva – the insect with a three-stage life, or the insect with a four-stage life? (*The insect with a three-stage life*)
6 Which is the only stage of life in which an insect can have wings? (*The adult stage*)
7 How many wings have most adult insects? (*Four*)
8 How many feelers do insects have? (*Two*)
9 Which insects are called caterpillars in the larval stage? (*Moths and butterflies*)
10 How many true legs does a caterpillar have? (*Six*)

## Relevant Information

The purpose of this lesson is to observe the main characteristics of those insects which are classed as moths and butterflies.

Both wings and bodies of adult moths and butterflies (*Lepidoptera* – scale wings) are covered with very tiny powdery scale-like particles. This fine 'dust' is responsible for their delicate colouring.

### Eggs

The eggs of some moths and butterflies are oval in shape; others are flattened in appearance, spherical, or hemispherical. The shells of these eggs are usually ribbed or sculptured in some way. The number laid is variable, and they are usually laid on or near the particular food on which the emerging larvae will feed.

EGGS OF MOTHS        EGGS OF BUTTERFLIES

Oak egger    Puss    Vapourer      Comma    Milkweed    Small blue

## Larvae

The larva of any moth or butterfly is generally called a caterpillar. Its body is divided into thirteen segments, the first three of which form the thorax and carry a pair of legs each. The remaining ten form the abdomen, and here the last two segments are not always easily distinguished from one another. It is on the sixth to ninth of the segments of the abdomen that the pro-legs – known as false legs or stumps – are to be found (except on most looper caterpillars). The separate pair of pro-legs on the final segment are referred to as the anal claspers.

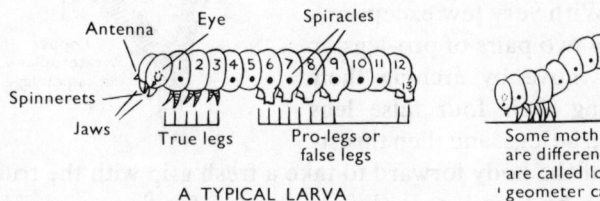

A TYPICAL LARVA

Some moth caterpillars are different. They are called loopers or ' geometer caterpillars

The nine pairs of spiracles or air holes which are used for breathing look like small dots along the sides of the body and are quite often easily seen.

Although the bodies of some caterpillars are quite smooth, others – e.g. the garden tiger moth larvae – have hairy tufts growing from the sides, and these serve as protection. Children should be discouraged from over-handling such caterpillars, because of the possible effects of skin irritation.

The larvae of all the lepidoptera (moths and butterflies) have two spinnerets in the head, from which comes the liquid which, on drying out, forms the familiar silken threads. The spinning glands from which this liquid comes are modified salivary glands. In the case of the so-called silk-worm, these glands are about five times as long as the whole of the caterpillar, and from these larvae comes the silk used for commercial purposes. Caterpillars in general use the silken threads which they manufacture for a variety of purposes such as:

a attaching materials such as bits of bark and leaves to provide shelter
b constructing ropes for climbing purposes

*c* constructing communal webs within which the caterpillars feed
   for the first part of their existence as larvae
*d* providing a protective envelope for the pupa
*e* fixing or suspending the pupa.

Looper caterpillars are the larvae of moths belonging to the sub-
family known as *geometridae*. There are some 270 species in Great
Britain, and British entomologists refer to them as geometers or
loopers, but Americans refer to them as measuring worms or span-
worms. The appearance and attitude
adopted by some of them provides
excellent camouflage and has resulted
in their being occasionally mis-named
stick insects. With very few exceptions
they have only two pairs of pro-legs or
claspers, and move by arching their
bodies bringing their four false legs
up to their six true legs, and then thrust-

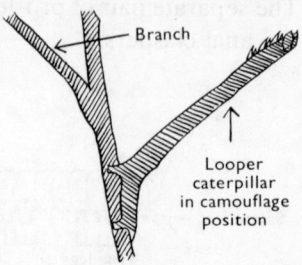

ing the front of the body forward to take a fresh grip with the true
legs. They may be found, according to species, feeding on a wide
variety of herbs, trees and shrubs.

    Caterpillars have mouth parts adapted for biting and chewing.
The vast majority feed on some part of a herb, tree or shrub –
usually the leaves. Living roots, stems, leaves, flowers, fruits and

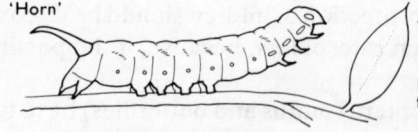

The 'horn' is a characteristic of the various Hawk Moth caterpillars, eg Poplar Hawk, Elephant Hawk, Privet Hawk, Lime Hawk etc. They look ferocious but are harmless.

seeds all serve as food, however, although caterpillars do not seem
to be found on coniferous trees, and indeed very rarely on any of the
evergreen trees and shrubs in the British Isles. Some feed on dead
leaves, and some are timber-borers. In many cases, a species has its
own particular food plant and no other. Some are pests in stored
foods, *e.g.* flour moths form colonies in stored flour and corn.

    Very few feed on animal foods. One or two species of moths and
butterflies are cannibals, and the larvae of certain small moths feed
on dead animal parts such as feathers, leather, fur, wool, and other
mammal hairs.

## Pupae

The pupa of a moth or butterfly is generally called a chrysalis. In the majority of species there is little or no power of movement during the pupal stage, and the pupae are covered by a continuous hard coating. In many species the caterpillars, before pupation, surround themselves with a silken cocoon. In some cases the cocoon is attached to, or suspended from, some suitable object by silken threads. There are a few species of moths where the pupal stage is reputed to last for four, five, or even six years.

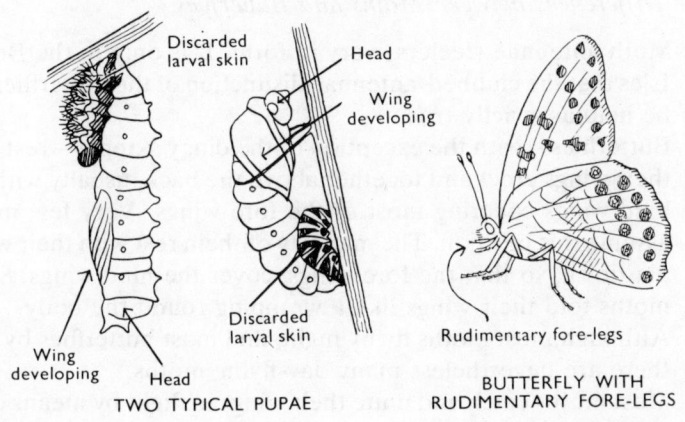

Discarded larval skin — Head — Wing developing — Wing developing — Head — Discarded larval skin — Rudimentary fore-legs

TWO TYPICAL PUPAE

BUTTERFLY WITH RUDIMENTARY FORE-LEGS

## Adults

Among moths and butterflies the adult males have four wings, and in almost all species the adult females also have four wings. There are a few small moths whose wings are rudimentary and useless.

Certain adult butterflies, such as the small tortoise-shell, make little use of the front pair of legs, which have become rudimentary. They stand on their other four legs, and this may lead to the careless conclusion that they have only four. Closer examination will reveal the other pair on the front segment of the thorax.

The mouth parts of adults – except for some extremely tiny lepidoptera – are adapted for sucking liquids only. Some moths have no mouth parts at all in the adult stage, e.g. the eggers, lappets, emperors, swifts and goats. Herbs, trees and shrubs which have young by means of flowers, fruits and seeds provide food not only for the majority of caterpillars, but also for adult moths and butterflies. Nectar from flowers is the chief food, but in some cases the juices of over-ripe fruits are taken.

Apart from the polar regions, the distribution of moths and butterflies is world-wide – almost wherever there are flowering plants. There must be over 100 000 species of moths throughout the world, and over 2000 of these are found in the British Isles. There are some 13 000 known species of butterflies throughout the world, but those which are accepted as British, number only about 68.

## The Differences between Moths and Butterflies

1   Moth antennae (feelers) vary in form. It is only in the British Isles that the clubbed-antennae distinction of the butterflies can be held as strictly true.
2   Butterflies – with the exception of the dingy skipper – rest with their wings erect and together above the back, usually with the hind wings covering most of the fore wings. Very few moths assume this position. The majority of them rest with their wings flat down, so that the fore wings, cover the hind wings. Some moths fold their wings like a wrapping round the body.
3   Although most moths fly by night, and most butterflies by day, there are nevertheless many day-flying moths.
4   Most moths catch and unite their wings in flight by means of an underwing bristle, although some – for example, the emperors and eggers – lack these bristles. No butterfly possesses this bristle.
5   Butterflies are generally accepted as being the 'brightly coloured ones', but although it may be fair to say that the dullest colours are found among moths, there are nevertheless many very brightly-coloured moths.
6   All British butterfly larvae have ten pro-legs (including the anal claspers). Many moth caterpillars also have ten pro-legs, but some are without this number.
7   British butterfly pupae are usually found suspended and tied with silken thread, and sometimes lightly cocooned. British moth pupae are rarely suspended, and sometimes fully cocooned; others are laid underground without any cocoon.

As a general rule the British moths and butterflies complete their life span during one year. Again generally speaking, this may be divided into four periods.

248

*a* Egg   – one month
*b* Larva – two months
*c* Pupa  – two weeks
*d* Adult – one month

To this may be added a period of about six months' winter sleep. Some species spend the winter as eggs, some as larvae, some as pupae, and some as adults. There are some short-lived species, however, which have up to as many as four generations in one year, the winter generation living the longest. Others – especially those whose larvae are wood-borers – may have a larval development lasting two, three or four years.

The illustrations for Lesson 28 in the Pupils' Book show:

*Large White or Cabbage Butterfly*. This is a common butterfly. The female is distinguished from the male by two black spots on the fore wings, and a black dash on the margins between the wings. Eggs are laid in batches of up to one hundred. The larvae, which are notorious for their offensive smell, feed on cabbage leaves. The pupae are often found in sheltered corners of walls and fences.

*Garden Tiger Moth*. The adults, which are night flyers from June to August, are beautifully marked. The larva, which is a great favourite with children, is known by them as the 'woolly bear'. It feeds on a wide variety of plants including dock, nettle, and various shrubs. The bristles of the body are used during the formation of the cocoon.

*Red Admiral Butterfly*. The adults, which feed on nectar and fruit juices, are on the wing from June to late October. Eggs are laid on nettle plants, and each larva makes itself a kind of tent by fastening nettle leaves together with silk. The larval stage lasts about a month, and that of the pupa, which is suspended, about two weeks.

*Puss Moth*. The adult, with its fluffy body, is on the wing at night during May, June and July. The tawny-brown eggs are usually laid in pairs on poplar or willow, on which the larvae later feed. When frightened, the larva raises its two tails like stings and curves them forward over its back. For the pupal stage a cocoon is spun and attached to some part of the tree.

## CODE

1   Observe the main difference between the feelers of an adult British moth and those of an adult British butterfly.
2   Observe the position of the wings of an adult British butterfly at rest, and of an adult British moth at rest.
3   Observe – when caterpillars are discovered – not only the caterpillars but the type of plant on which they are feeding.
4   Observe the different varieties of caterpillars which may be feeding on one kind of plant, especially trees.
5   Collect caterpillars from a tree or shrub, and rear in an insect cage – together with their food plant. (See Lesson 6.) Once they pupate, they should be left undisturbed until the adults appear.
6   Observe the silken threads spun by caterpillars, and for what purpose they are used.
7   Observe any captive moths or butterflies in as large an insect cage as possible. In the absence of suitable flowers, food may be provided in the form of a cotton wool pad soaked in a solution of honey and water, or a syrupy solution of sugar and water. A small paste pot or ink bottle may be used to hold the pad.
8   Observe any acquired flour moths – sometimes found in stored forms of flour or corn. Successive generations may be reared for years in a jar of Bemax or flour, with no attention other than the occasional addition of more food. (See Lesson 6)
9   Preserve eggs, larvae and pupae in 5% formaldehyde in the usual way.

### Note
Mounted adults may be obtained from biological suppliers. In such cases, the wings of butterflies will be displayed horizontally, unless otherwise requested.

### Written Work

1   Adult moths and butterflies have powdered wings.
2   The larva is called a caterpillar. It has six jointed legs and a mouth that can bite and chew.
3   The pupa of a moth or butterfly is called a chrysalis.
4   The adult butterfly has knobs on its feelers.
5   The adult moth rests with its wings folded down.

## THE TWO-WINGED FLIES

### Demonstration Material

1   Any living or preserved examples of:
    a the eggs of a two-winged fly, *e.g.* bluebottle eggs on a piece of raw meat
    b the larvae of a two-winged fly, *e.g.* maggots, crane fly larvae ('leatherjackets'), gnat larvae, midge larvae
    c the pupae of a two-winged fly, *e.g.* bluebottle pupae, gnat pupae.
2   Any living or mounted example of an adult two-winged fly *e.g.* crane fly, bluebottle, housefly.

### Sample Link Questions

1   What are the stages in the life of an insect with a four-stage life? (*Egg, larva, pupa, adult*)
2   Which looks like the adult – the larva of an insect with a four-stage life, or the larva of an insect with a three-stage life? (*The larva of an insect with a three-stage life*)
3   How many jointed legs has the larva of a moth or butterfly? (*Six*)
4   What special name do we give to the larva of a moth or butterfly? (*Caterpillar*)
5   During which stage do most insects with a four-stage life rest? (*The pupal stage*)
6   What special name do we give to the pupa of a moth or butterfly? (*Chrysalis*)
7   How many wings has an adult moth or butterfly? (*Four*)
8   What gives the wings of a moth or butterfly their colour? (*Very tiny powdery scales*)
9   What are the two main differences between an adult British moth and an adult British butterfly? (*The butterfly has knobs on its feelers. The moth has not. The butterfly rests with its wings folded up; the moth rests with its wings folded down*)
10  Have all adult insects four wings? (*No, some have two, some none*)

## Relevant Information

The purpose of this lesson is to observe the main characteristics of those insects which are classed as two-winged flies.

The characteristics of beetles and of bugs are the subjects of separate lessons in Book 3.

The characteristics of dragonflies and caddis flies (which are not true flies) are the subject of a separate lesson in Book 4.

Of the true flies (*Diptera* – two wings), there are well over 50 000 species in the world, and in the British Isles up to about 5000 species. These insects are distinguished from others by having only the front pair of wings for flying purposes. The hind pair have been reduced and are referred to as balancers or halteres. There are a few exceptions to this rule, as some of the very small flies are completely wingless.

### *Eggs*

The eggs of the true flies are of various forms and are usually laid in large numbers. A single housefly, for example, may deposit some 2000 eggs in a lifetime, in batches of about 75 to 150. Some flies lay their eggs in loose masses, and others in rows. Parasitic flies usually lay their eggs on or in other insects. The egg stage is usually short, varying from a few hours to a few weeks.

### *Larvae*

The larvae of the two-winged flies have a variety of forms, some of which are referred to as maggots. Mostly they are tubular in shape, and very often taper towards the head end. On many the head itself is considerably reduced and may not be visible. Although some kinds of larvae have fleshy stumps, none have jointed legs. This obviously distinguishes them from the larvae of moths and butterflies.

Larvae vary in their habits, and the different species enjoy a wide range of foods.

a Living plants – roots, stems and leaves.
b Living animals – some larvae live and feed on other animals as external parasites, and others live and feed as internal parasites.
c Dead animal parts.
d Dead plant parts.
e Various excrements.

Two examples of larvae          Immobile and mobile pupae

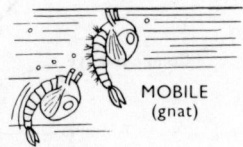

With no visible
head (bluebottle)

With visible
head (gnat)

IMMOBILE
(housefly)

MOBILE
(gnat)

## Pupae

The pupae of some of the two-winged flies are enclosed by the final larval skin, which becomes a hardened shell or puparium. The pupae of others are free from this final skin. In some species the pupa is immobile, *e.g.* the housefly; in other species it is mobile, *e.g.* the mosquito.

## Adults

In the adult stage the true flies have only one pair of wings. The three segments of the thorax are fused together, making it difficult to see each segment separately. The head is usually large and dominated by a pair of large compound eyes with hundreds of facets. Between these compound eyes, there are often three simple eyes. The mouth parts of the adults are adapted for sucking, and sometimes in certain females also for piercing. These mouth parts form a trunk or proboscis.

Adult flies in general feed on flower nectar or scavenge on sweet or putrid liquid matter. Females of certain species – such as the mosquitoes (of which group the gnat is a member), the midges, and horseflies – have mouth parts adapted for piercing as well as for sucking. These are used for obtaining mammal blood which is apparently a necessary form of nourishment prior to egg-laying. Blood-sucking habits are confined to the females in the fly family, with the exception of the very small parasitical pupipara. Amongst these insects both sexes are blood-suckers.

In most of the common flies there may be several generations in a year. Male flies often die shortly after mating, and female flies after egg-laying.

Some flies spend the winter in the egg stage, some in the larval stage, some in the pupal stage, and others in the adult stage.

Although most flies, like most other insects, have young by means of eggs, there are some which retain their eggs until after they have hatched, and produce living larvae instead. Certain of the flesh flies,

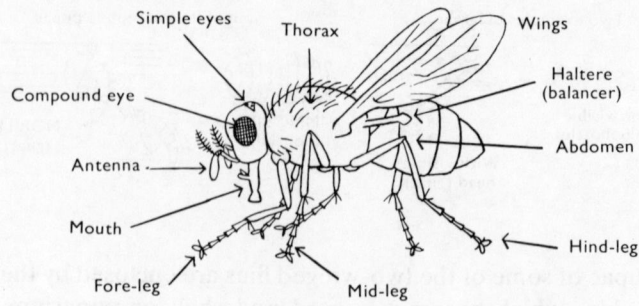

PARTS OF A TYPICAL ADULT DIPTERA (TRUE FLY)

for example, can deposit living larvae small enough to creep through the gauze of a meat safe. The tsetse flies and the parasitic louse flies are able to retain and nourish their larvae almost up to the stage of pupation, but these are exceptional.

The two-winged flies are both useful and harmful to other living things. Those which are scavengers obviously serve a useful purpose. A number of them also assist in pollination. Various animals also use adult flies and their larvae for food. Such animals include poultry and fish and, occasionally, certain primitive races of human beings.

Blood-sucking flies and flies whose larvae infest open wounds are a particular nuisance to some animals, and in addition there are some which are agents in spreading various diseases. These include the common housefly, which carries the germs of several diseases such as typhoid fever and infant cholera, mosquitoes which transmit malaria and yellow fever, and the tsetse fly which carries African sleeping sickness.

Certain insects which are termed flies are not really flies at all. For example, the caddis flies, the mayflies and the dragonflies are all four-winged insects and are therefore not true flies.

The illustrations for Lesson 29 in the Pupils' Book show:

*Housefly.* Eggs numbering 100 to 150 are laid in mammal dung, on which the larvae feed. The larvae, which emerge within about ten hours, are sightless and legless. After about a week they descend into the ground for about an inch, and pupate in hard cases. The adult appears from the pupal skin within a week to a month, depending on temperature, and takes the same food as the larva. Its unsavoury diet results in its being a carrier of diseases.

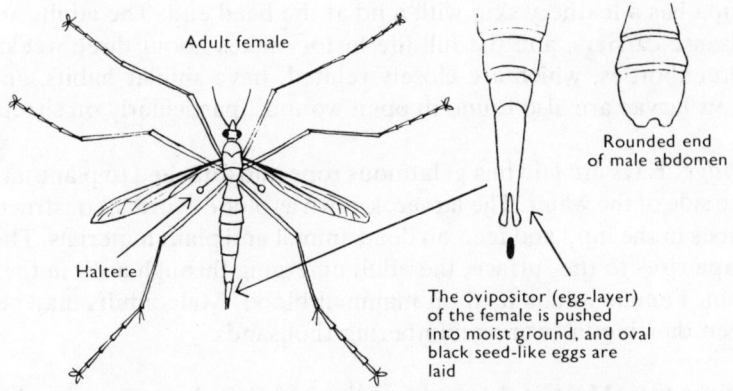

Adult female

Rounded end
of male abdomen

Haltere

The ovipositor (egg-layer)
of the female is pushed
into moist ground, and oval
black seed-like eggs are
laid

DIFFERENCE IN APPEARANCE OF FEMALE AND MALE CRANE FLIES

*Crane fly (daddy-long-legs)*. Eggs are laid among the roots of grasses, the abdomen of the female being pointed for this purpose. The eggs hatch in about a week, and the legless, tough-skinned, brownish larvae are known as leatherjackets. They feed on grass roots and prefer damp conditions. Pupation takes place after about a year; the cylindrical pupa is brown or greyish. Adults are clumsy fliers. The male is distinguished from the female by having a rounded tip to its abdomen.

*Gnat (Culex)* – one of the mosquito group. 200 or more eggs are laid in an unsinkable raft on water. The larva feeds on dead animal or dead plant material in the water and breathes at the surface through a tail tube. The pupa is active but does not feed. It breathes at the surface through two tubes on its head. The adult emerges from a slit between the breathing tubes and feeds by sucking the dew from plants. Females add mammal blood to this diet. The illustration in the Pupils' Book shows a female adult gnat with its mouth part puncturing the skin of a mammal for this purpose. The malaria-carrying mosquito transmits the disease in this way. Development from egg to adult takes about a month and can occur in a very small amount of water.

*Bluebottle*. Our largest (the emetic beauty bearer) is the one commonly found near dwelling places. The long whitish eggs are laid on carrion, fish and meat. The white sightless larvae, known as maggots or gentles, dissolve their food and suck up the resulting liquid. The pupa has a leathery skin with a lid at the head end. The adults are disease carriers, and the full life history takes about three weeks. Greenbottles, which are closely related, have similar habits, and their larvae are also found in open wounds, particularly on sheep.

*Midge*. Eggs are laid in a gelatinous rope and attached to plants at the side of the water. The larvae, known as blood-worms, construct tubes in the mud and feed on dead animal and plant materials. The pupa rises to the surface, the adult emerging through a slit in the skin. Female adults feed on mammal blood. Male adults may be seen dancing in swarms numbering thousands.

*Hover flies*. Most of the adults of the 230 British species resemble bees or wasps in colouring and in their humming noise when in flight. They are distinguished outwardly from bees and wasps by having only one pair of wings, and by their lack of a waist. Flight is swift and darting, and they pause to hover almost stationary in the air. The food of the larvae varies according to species – those which feed on aphids, those which feed on the refuse from the nest of a wasp or bee, those which feed on dung or dead plants. Certain of the last type are aquatic and are known as rat-tailed maggots because of their rear telescopic breathing tube. This enables the larva to obtain its oxygen from the air and may be extended or retracted at will.

*Horse-flies*. These are also known as gadflies or breezeflies, and they include some of the largest British flies. Eggs are usually laid on swamp plants. The larvae feed on other insect larvae or worms. The pupae are long and cylindrical. The males suck flower nectar, honey dew or even water, but female adults pierce mammal skin to obtain blood and are painfully persistent in this respect.

## CODE

1  Observe the single pair of wings on any two-winged fly.
2  Observe the balancers on an adult fly and their position. These are easily observed on a resting crane fly.
3  Observe the sucking mouth parts (proboscis) on any adult two-winged fly, particularly the bluebottle and housefly.
4  Observe that on the larva of any two-winged fly there are no legs, *e.g.* maggot.

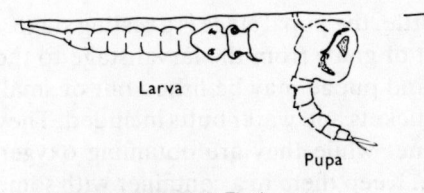

Larva

Pupa

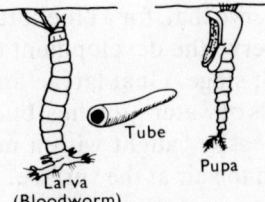

Tube

Larva
(Bloodworm)

Pupa

Mosquito (*Anopheles*). The larva is distinguished from that of the gnat by its very short breathing tube, and by its lying horizontally at the surface. Occurs mainly in Southern England and the Fen districts.

*Chironomous* Midge. The crimson larvae are well-known as 'bloodworms'. There are several species, some constructing tubes in the mud. Often scooped up with mud. The pupae may be observed jerking their way with stiff-jointed undulations through the water.

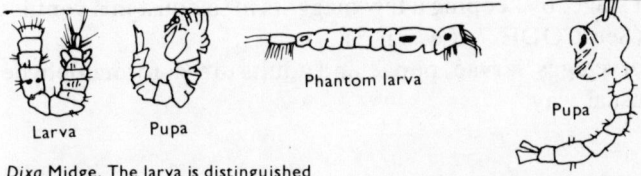

Phantom larva

Larva    Pupa

Pupa

*Dixa* Midge. The larva is distinguished by being characteristically bent double and takes in oxygen through tail openings

*Chaoborus* Midge. The larva is very transparent and termed the Phantom midge accordingly. Carnivorous, it normally lies motionless and horizontal. Clearly visible are the black eyes, and two air sacs, one at each end of the animal's body.

The larvae and pupae of gnats, mosquitoes and midges may be housed for observation purposes in a toffee jar with a perforated screw top, and should require no attention.

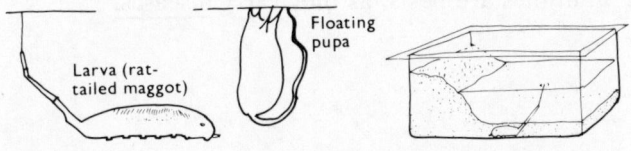

Floating pupa

Larva (rat-tailed maggot)

Hover fly or Drone fly (*Eristalis*). The larva—known as the Rat-tailed maggot, buries itself in the mud and obtains oxygen from the air by means of a long telescopic siphon, which it can extend to several times the length of its own body. They may be collected and kept in a container together with some of the mud and pond water from where they were found. If some kind of soil and grass bank is built up in their container, they may pupate and pass through to the adult stage. A cover of butter muslin will prevent the immediate escape of the adults.

SOME EXAMPLES OF AQUATIC LARVAE AND PUPAE

257

5 Observe that, for a bluebottle, the pupal stage is a resting stage.

6 Observe the development of gnats from the larval stage to the adult stage. Gnat larvae and pupae may be fished out of small pools of water – ditches, buckets and water butts included. They are easily caught with a net while they are obtaining oxygen from the air at the surface. Keep them in a container with some of the mud from the water in which they were found. As the complete life cycle lasts about a month, it is easy to have larvae, pupae, pupal skins and adults all present in the one container at the same time.

7 Observe that a gnat pupa is free-swimming in the water.

8 Observe the aquatic larvae or pupae of any other two-winged flies commonly found in freshwater ponds and ditches, etc.

9 Observe the development of bluebottles from the larval to the adult stage, by keeping a few maggots in sawdust in a ventilated jar. (See CODE 7, Lesson 5).

10 Preserve eggs, larvae, pupae and adults in 5% formaldehyde in the usual way.

*Note*

Mounted adults if required, may be obtained from biological suppliers.

**Written Work**

1 A true fly has only <u>two</u> wings for flying.
2 The <u>larva</u> may be called a maggot.
3 There are <u>no</u> legs in the larval <u>stage</u>.
4 Some pupae <u>rest</u>. Some move from <u>place</u> to <u>place</u>.
5 Some adult flies are pests, as they carry <u>diseases</u>.

# 30  NEVER-ALIVE THINGS IN SPACE

## STARS, PLANETS AND SATELLITES

**Demonstration Material**

1  Three children.
2  Models to show relative sizes of the first four planets and their satellites. (See CODE 2.)

**Sample Link Questions**

1  What is the sun? (*A star*)
2  Is every star smaller than the sun? (*No, some are bigger*)
3  Why do all stars seem smaller than the sun? (*Because they are so very far away*)
4  Why do we not see the stars in daylight? (*Because the sun outshines them*)
5  Which is bigger – the earth or the sun? (*The sun*)
6  Why does the sun seem smaller? (*Because it is so far away*)
7  Which is bigger – the earth or the moon? (*The earth*)
8  Does the sun go round the earth, or the earth round the sun? (*The earth goes round the sun*)
9  How long does it take the earth to go once round the sun? (*One year*)
10  Does the earth go round the moon, or the moon round the earth? (*The moon goes round the earth*)
11  About how long does it take the moon to go once round the earth? (*A month*)

**Relevant Information**

The main points of a lesson in Pupils' Book 1 on the earth and the moon were:

1  the earth is bigger than the moon
2  the moon goes round the earth
3  it takes a month (lunar) to go once round the earth.

The main points of a lesson in Pupils' Book 1 on the earth and the stars were:

259

1   the sun is a star
2   the earth goes round the sun
3   it takes a year to go once round the sun.

The main points of a lesson in Pupils' Book 1 on day and night were:

1   we have daylight and night-time because the earth spins
2   the earth takes a day to spin round once.

The main purpose of this present lesson is to show the difference between a star, a planet, and a satellite:

1   the stars that we see consist of glowing gases
2   a planet goes round a star
3   a satellite goes round a planet.

The solar system is the subject of a lesson in Pupils' Book 3.

Stars have a separate lesson in Book 4.

### Stars

Stars are considered to be spheres of glowing gases. They do not burn in the way that a fire burns. They glow as the result of heat, in the way that a red-hot poker glows. Their heat is such that even their metals are expected to be in the form of a gas.

The sun is the nearest star to the earth. Its distance is approximately 150 million kilometres, but this is short compared with the distance of other stars. The next nearest is a very faint star – Proxima Centauri at a distance of about 4.25 light years, or some 40 000 000 000 000 kilometres. A space ship travelling at 40 000 kilometres per hour would take over 114 000 years to reach it.

It is the sun's relative nearness which makes it look larger and brighter than any of the other stars. Some stars are brighter than the sun, and some are less bright. The brightest known star – Sigma Doradus – is calculated to be more than 300 000 times brighter than the sun, although its considerable distance makes it appear very faint. Sirius, the dog star, which appears to be the brightest star in the sky is only 26 times as bright as the sun, but is much nearer to us than Sigma Doradus – a mere 80 million million kilometres away. It is still far enough away for its light to be completely outshone by the sun during daytime. Some stars are hotter than the sun, and some are cooler. The coolest of the stars – the dark stars – are not visible to us.

Despite the vast spaces between one star and another, they are nevertheless in groups, known as galaxies. These galaxies are in turn in groups. Vast gulfs of space separate one group of galaxies from another group of galaxies. The sun is only a star of average size and brilliance in our own particular galaxy, which consists of some 100 000 million stars. Our galaxy is not really of any special importance in the interminable universe.

Although the vast number of stars in a galaxy and the vast number of galaxies make it impossible to calculate the number of stars in existence, there are only about 6 000 visible to the unaided eye. A telescope is required to see more.

Shooting stars are not stars at all. They are meteors attracted by the earth's gravity. As they fall towards the earth, they are vaporized in the atmosphere by friction, and this results in the streak of light which we see.

## Planets

Planets go round a star. The sun is the only star which we know with certainty to have planets going round it; other stars are too far away for us to be certain at present. It should not be assumed that out of all the countless billions of stars in the universe, the sun is the only one that can have planets. In fact, if only one star in 100 000 had planets, that would still leave a million planetary systems in our own galaxy alone. It has been discovered that some of the nearer stars have non-luminous companions, some smaller, and others larger, than our local planet Jupiter. These are now regarded as being stellar planets.

## Satellites

Satellites go round a planet. In the solar system, Mercury and Venus are the only planets which, so far as is known, are without satellites. The earth has one natural satellite, and whatever artificial satellites are in orbit at the time of reading. Pluto has one satellite; all the other planets in our system have more than one.

*The largest known stars* are inconceivably massive. Typical of these is IRS 5. It is a cold giant star with a diameter of some 15 000 million kilometres. The diameter of the sun is 1 400 000 kilometres. This means that more than 10 000 stars the size of the sun could be placed in line along its diameter. In fact, if the sun could be placed at its centre, there would be plenty of room for all of the planets to revolve in their orbits round the sun. Larger stars than this are being discovered.

*The smallest stars* are smaller than the moon. Typical of these is LP 327–186. It is about 100 light years away, and its diameter is only half that of the moon. Smaller stars than this are being discovered.

*The largest planet* in the solar system is Jupiter. It has a diameter of about 142 800 kilometres. The diameter of the earth is about 12 700 kilometres, which means that eleven planets the size of the earth could be placed in line along the diameter of Jupiter. 1 300 earths could fit inside it.

*The smallest planet* in the solar system is Mercury, the nearest planet to the sun, with a diameter of about 4 900 kilometres.

*The largest satellite* in the solar system is as yet to be accurately determined. It was at one time believed to be Titan, one of the satellites of Saturn, but Ganymede and Callisto, two of the satellites of Jupiter are very similar in size, and may be larger. All three are roughly the size of the planet Mercury.

*The smallest satellites* (not counting man-made satellites) are thought to be Deimos (of Mars) and Leda (of Jupiter), each having an estimated diameter of less than 14 kilometres.


### Our Own Star, Its Planets, and Their Satellites

To our present knowledge the number of planets revolving round the sun, is nine. The orbits of these planets are elliptical, and roughly in the same plane as the sun's equator. If viewed from above the earth's north pole all the planets, including the earth, would appear to revolve round the sun in an anti-clockwise direction. Speed varies, as the planets travel over the part of their orbit which is nearer to the sun at a faster rate than they do over the part of the orbit which is farthest away. The time taken to complete one orbit round the sun also varies from planet to planet; the greater the orbit, *i.e.* the greater the distance to be travelled, the longer it takes to go round the sun. Thus Mercury, the nearest planet to the sun takes only 88 days (earth-time) to complete one orbit, whilst the farthest planet, Pluto, takes almost 248 years.

## The Sun

The sun, like other stars, is considered to be a sphere of glowing gases. It is NOT a ball of fire as is sometimes stated. Hydrogen seems to be the main gas in its composition. The temperature at its surface is estimated to be about 6 000°Celsius, and the temperature at the centre in excess of 15 000 000°Celsius. The diameter of this, our parent star, is some 1 392 000 kilometres (1 400 000 kilometres approximately), or about 3½ times the distance of the earth from the moon, which means that over 109 planets the size of the earth could be placed along its diameter, and over 1 000 000 planets the size of earth could be packed within. If earth were placed at its centre, there would be enough room for the moon in its orbit round the earth. The sun spins on an axis, taking 25 380 days to complete one rotation. Its gravitational pull holds the planets in their orbits and prevents them from escaping into outer space, in the same way as the satellites are held in orbit round their parent planets.

| Planets (in order of distance from the sun) | Average distance from the sun (in millions of kilometres) | Number of times the earth's distance from sun (approx) | Diameter at the equator (in kilometres approx) | Number of times the earth's diameter (approx) | Time taken to complete one orbit round the sun (in earthtime) | Number of satellites known |
|---|---|---|---|---|---|---|
| Mercury | 58 | 0.4 | 4 900 | 0.4 | 88 days | 0 |
| Venus | 108 | 0.7 | 12 400 | 0.96 | 225 days | 0 |
| Earth | 150 | 1 | 12 750 | 1 | 1 year | 1 |
| Mars | 228 | 1.5 | 6 760 | 0.55 | 687 days | 2 |
| Jupiter | 779 | 5 | 142 800 | 11 | 11.9 years | 16 |
| Saturn | 1,426 | 9.5 | 120 900 | 9 | 29.5 years | 16 |
| Uranus | 2,870 | 19 | 48 000 | 4 | 84 years | 5 |
| Neptune | 4,493 | 30 | 50 000 | 4 | 164.8 years | 2 |
| Pluto | 5,900 | 40 | possibly 6 000 | 0.5 ? | 247.7 years | 1 |

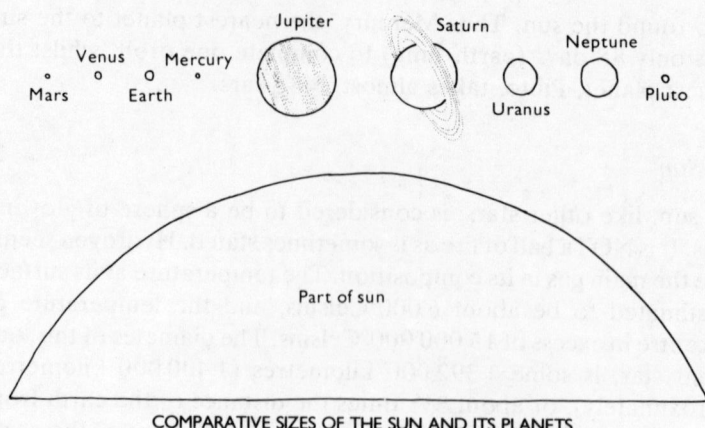

COMPARATIVE SIZES OF THE SUN AND ITS PLANETS

## Notes

*i* Neptune's diameter is based on a 1969 calculation. Prior to that, it was believed to be slightly less than that of Uranus.

*ii* The data for Pluto is still subject to doubt.

## The Planet Mercury

This, the smallest planet, and the nearest to the sun, completes one revolution round the sun in 88 days (earthtime). It is now known to rotate very slowly on its axis in the same direction as it revolves round the sun, *i.e.* in the same direction as the earth. This rotation is very slow – once in every 58½ earth days. The 1974 probe found its surface to be something like that of the moon, and littered with many craters, ranging in size from 1 kilometre to over 200 kilometres across. During its slow rotation, this chaotic terrain is alternately baked by the sun and frozen by the bitter cold of space. There is a small magnetic field, and the thin atmosphere which is present contains hydrogen, helium, neon and argon, and is probably lacking water vapour or free oxygen, so that no form of life as we know it is likely there.

## The Planet Venus

Venus, very nearly the size of the earth, is seen – like the other planets and satellites – only because it reflects light from the sun. Even so, it appears as the brightest thing in the sky apart from the sun and moon. Even when the giant planet Jupiter is visible, Venus still appears brighter. When circumstances permit, it can be seen in the sky for about four hours after sunset, and for about four hours before sunrise, and is the bright 'star' seen shining alone or in the presence of the moon in the evening sky. Because of this, it is often called the morning star or the evening star. Venus orbits the sun once every 225 earth days. It is now known to rotate slowly on its axis, in the reverse direction to that of the earth, about once every 243 earth days so that, to a Venusian, the sun would appear to rise in the west and to set in the east. The rotation is such that, to an observer on Venus, a complete day from sunrise to sunrise would last about a Venusian year.

Although the surface of Venus is perpetually obscured by a cloudy atmosphere, this was eventually penetrated by Russian probe vehicles which revealed the surface to be intolerably hot, and the atmosphere to consist of about 97% carbon dioxide and about 2% nitrogen, with only traces of oxygen and water vapour, so that no form of life as we know it is likely to exist there.

### The Planet Earth

The diameter is about 12 760 kilometres, and the circumference some 40 000 kilometres. It rotates once on its axis in 24 hours, turning from west to east, *i.e.* anti-clockwise to an observer in the northern hemisphere. Its speed of rotation is about 1 600 kilometres per hour at the equator, and about 1 100 kilometres per hour in the temperate zone. During one revolution round the sun – a year – the earth completes 365¼ of these rotations, which is why we count 365 days in a normal year and 366 days every fourth year – the leap years.

In its elliptical orbit round the sun, the earth travels approximately 940 million kilometres each year at an average speed through space of roughly 107 000 kilometres per hour (about 30 kilometres per second).

### Earth's Natural Satellite, the Moon

The diameter is about 3 470 kilometres – roughly a quarter of that of the earth – and it would take almost 50 moons to equal the volume of the earth. Its weight is about one-eightieth that of the earth, and its force of gravity about one-sixth that of the earth. The moon revolves around the earth in an anti-clockwise direction (from west to east), this being the same direction as that of the earth around the sun. However, owing to the earth spinning faster than the moon goes round it, the moon appears to rise in the east and set in the west as does the sun. It rises about 50 minutes later each day, so that for about two weeks it rises at night-time, and then for about two weeks during daylight.

The orbit of the moon round the earth is not circular, but elliptical, so that its distance from the earth varies from about 348 000 kilometres to about 399 000 kilometres. The average distance is about 376 000 kilometres which is less than ten times the distance round the earth's equator, or about thirty times the earth's diameter. The speed of the moon in its orbit is about 3 600 kilometres per hour. The actual time taken for the moon to travel once round the earth is 27 days 7 hours 43 minutes. (This is a sidereal month.) However, during this period, the earth itself moves on in its orbit round the sun, so that the time taken between the appearance of one new moon and the next is longer than this, *i.e.* 29 days 12 hours 44 minutes. (This is a lunar or synodic month.)

In addition to revolving round the earth, the moon also rotates on its axis in the same way as the earth does: but whereas it takes the earth a day, it takes the moon a month. It rotates only once, therefore, during one revolution round the earth. This results in the same face of the moon being always turned towards the earth, so that we never see the other side from the earth. However, commencing with the Russian probe Lunik 3 in October 1959, it has been well photographed from passing space vehicles. It is because of the slow rotation of the moon that a complete 'day' on that satellite would last about four weeks, and consist of about two weeks of light, and two weeks of darkness.

The surface is covered with craters, smooth plains and jagged mountains. The so-called seas (*maria*) are of course not seas at all, but relatively flat plains covered by solidified lava. Some of the craters are volcanic in origin, others were caused by the impact of meteorites falling from space without any atmosphere to restrict them. There are also cracks on the surface called 'rilles' which are like dried-up water courses. As there is no atmosphere of air to scatter the colour rays of light sent out by the sun, the 'sky' seen from the surface of the moon appears black. Because there is no atmosphere, there is no wind, no clouds, no weather, and no air for sound vibrations to travel through, so that communication on the moon has to be by radio. Although no evidence of free water has been found on the surface, there is water – minute quantities of it were extracted from rocks brought back by the Apollo 11 and 12 astronauts, and there still remain possibilities of underground deposits of ice.

## Man on the Moon (Brief notes on the Apollo missions)

1 July 20th, 1969: Apollo 11 module landed in Sea of Tranquillity with Neil Armstrong and Edwin Aldrin. On July 21st, Neil Armstrong became the first man to step out onto the surface.
2 November 19th, 1969: Apollo 12 module landed in Ocean of Storms with Charles Conrad and Alan Bean.
3 April 1970: The Apollo 13 mission to the Frau Mauro region was unlucky, and had to be aborted short of the moon due to technical failures. The astronauts were returned safely to earth.
4 February 5th, 1971: Apollo 14 module landed in Frau Mauro region with Alan Shepard and Edgar Mitchell.
5 July 30th, 1971: Apollo 15 module landed in Hadley Apennine region, with David Scott and James Irwin. During this and the two following landings, a lunar roving vehicle was used.
6 April 16th, 1972: Apollo 16 module landed in the Descartes region with John Young and Charles Duke.
7 December 11th, 1972: Apollo 17 module landed in Taurus Littrow region with Eugene Cernan and Dr. Harrison Schmitt, thus completing the Apollo programme.

## Earth's Man-Made Satellites

The first of these was launched by Russian scientists on 4th October 1957. It had a diameter of 58 centimetres, and weighed about 83 kilograms. Its maximum distance from the earth was about 1 000 kilometres, and its initial speed enabled it to circle the earth once in 1 hour 36 minutes. Since then, there have been literally thousands of artificial satellites launched of various shapes and sizes, and serving many different purposes. There are weather satellites, communication satellites, scientific observation satellites, military reconnaissance satellites, and navigation satellites. Skylab, illustrated in the Pupils' Book was designed by the Americans to be used as their first space station programmed for a long term orbital flight.

267

Launched in 1973, it contained a workshop, living quarters, and the necessary equipment to permit experiments in many technical fields, as well as docking facilities for vehicles arriving from, and returning to, earth. It disintegrated in July 1979.

## The Planet Mars

This, the red planet, is one of the most interesting planets in the solar system. When visible in the sky, it looks like a bright orange-red star. It rotates once on its axis in 24 hours 37 minutes 22 seconds, so that its 'day' is slightly longer than a day on earth. Information obtained from Russian and American space vehicles has shown Mars to be a cold, dry, dusty planet with a surface greatly altered by volcanic activity, erosion and crustal movement. There are relatively featureless plains, chaotic areas of jumbled ridges and valleys, and large areas covered by craters, possibly both volcanic and meteoric in origin like those of the moon. The once famous 'canals' do not exist as such, but are more probably long ridges of rock in deserts of dust, periodically revealed by the shifting storms of Mars. Also in the crust are great fissures (rilles), extending for hundreds of kilometres. These are possibly the dried-up beds of ancient water courses and one of them – some 4 800 kilometres long and 96 kilometres wide is even bigger than the Grand Canyon. The white polar caps which shrink in Martian summers and return in Martian winters are probably composed of frozen water. The thin tenuous atmosphere consists mainly of carbon dioxide with evidence of water vapour, and traces of oxygen, but in spite of this, and a low temperature, Mars is the most likely planet in the solar system – apart from Earth – where some form of life as we know it, might exist.

## The Satellites of Mars

The two dwarf satellites of Mars are irregularly shaped and densely cratered. Deimos, the smaller of the two, is about 23 500 kilometres from the planet, and completes an orbit round it once in 30 hours 18 minutes. It has an estimated diameter of 13 kilometres. Phobos, the larger of the two, is about 9 300 kilometres from Mars, and has a diameter of about 22 kilometres. It takes only 7 hours 39 minutes to complete one orbit round Mars, which means that it is travelling faster than Mars is spinning on its axis. This would give an illusion to

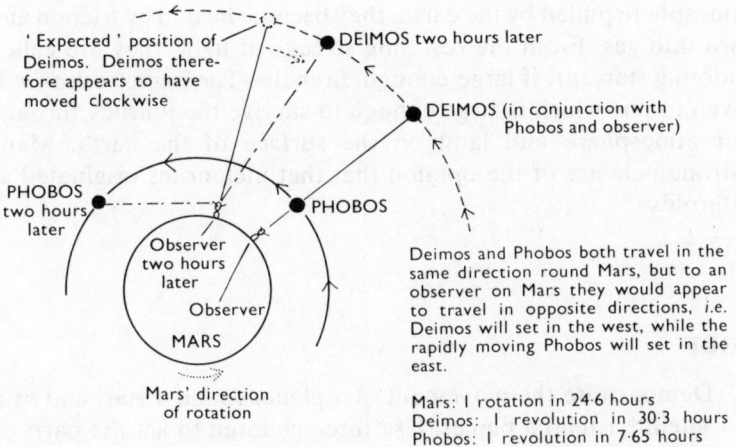

'Expected' position of Deimos. Deimos therefore appears to have moved clockwise

DEIMOS two hours later

DEIMOS (in conjunction with Phobos and observer)

PHOBOS two hours later

PHOBOS

Observer two hours later

Observer

MARS

Mars' direction of rotation

Deimos and Phobos both travel in the same direction round Mars, but to an observer on Mars they would appear to travel in opposite directions, i.e. Deimos will set in the west, while the rapidly moving Phobos will set in the east.

Mars: I rotation in 24·6 hours
Deimos: I revolution in 30·3 hours
Phobos: I revolution in 7·65 hours

an observer on Mars that Phobos and Deimos were travelling in opposite directions from one another.

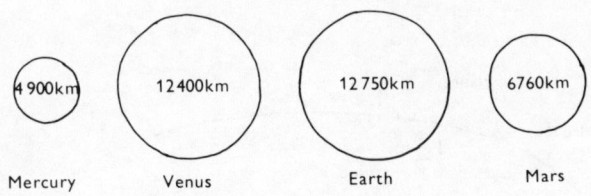

| 4 900km | 12 400km | 12 750km | 6760km |
|---|---|---|---|
| Mercury | Venus | Earth | Mars |

RELATIVE SIZES OF THE FOUR PLANETS NEAREST THE SUN

## Other Members of the Solar System (Asteroids, Comets and Meteors)

Most asteroids are in a belt which exists between the orbits of Mars and Jupiter. There are estimated to be over 40 000 of these minor planets or planetoids, all of them insignificant. According to one creditable theory, they are the remains of a shattered planet.

Comets go round the sun in long, elongated orbits. It is estimated that there are some 2 000 000 of them. Some, like Halley's comet, appear at regular intervals; others have orbits which take them so far into space that they may never return. A typical comet is made up of a solid core with a long tail of solid particles and tenuous gas.

Meteors are the dust specks of space. As they fall into our

atmosphere pulled by the earth, they become heated by friction and turn into gas. From the resulting streaks of light, they are called shooting stars, or, if large enough, fireballs. The name *meteorite* is given to one which is large enough to survive the journey through our atmosphere and land on the surface of the earth. Many astronomers are of the opinion that that meteorites originated as asteroids.

## CODE

1   Demonstrate the movement of a planet round a star, and of a satellite round a planet. Use three children to act the parts of star, planet and satellite:
  *a* star remains stationary
  *b* planet moves slowly round the star in a circle
  *c* satellite keeps with the planet, moving round it all the time.

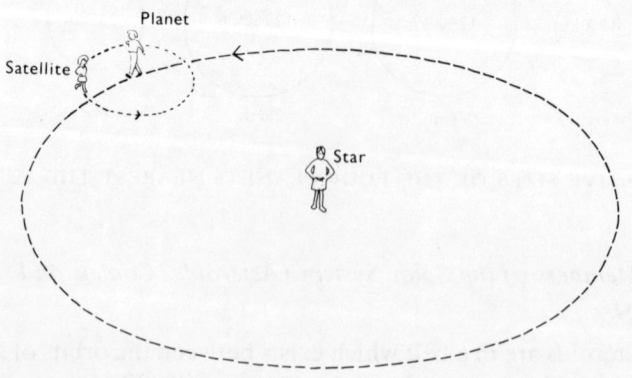

*Note*
The direction of movement in some other planetary system may be different, but in the solar system it is generally anti-clockwise (as viewed by an observer above the northern hemisphere of the earth). All the planets revolve round the sun in an anti-clockwise direction. With very few exceptions, all the satellites revolve round their

planets in an anti-clockwise direction. Rotation about an axis is likewise generally anti-clockwise (Venus is an exception). Even the sun itself rotates in an anti-clockwise direction, once every 25 380 days.

2   Demonstrate the approximate relative sizes of the first four planets of the sun, together with their satellites, by using the following scale models:

*a* MERCURY a ball of modelling clay, with a diameter of about 3 centimetres (bigger than moon, smaller than Mars).

*b* VENUS    a tennis ball – suggested colour yellow

*c* EARTH    a tennis ball – suggested colour green

*d* MOON     a glass marble

*e* MARS     a golf ball coloured red, or a slightly larger ball of red modelling clay

*f* PHOBOS   two pins stuck in modelling clay with tiny
AND          blobs of clay stuck on the heads.
DEIMOS

*Note*

  *i* Small rings of modelling clay could be used to stop the moon and planets rolling away; golfer's tee pegs stuck in modelling clay also serve as effective supports.

 *ii* It would not be feasible to demonstrate the relative distances from the sun on the same scale as that used for the models. On this scale too, the sun would present difficulties, as its model would need to have a diameter of 7 metres. However, the sun could be symbolised by a dot, and the relative distances illustrated by spacing the models along a line as shown in CODE 3.

3   Demonstrate the relative distances of the first four planets from the sun by spacing models along a line as shown in the columns below. It is emphasised that the scale of distances is different from the scale of sizes in CODE 2.

271

| PLANETS | DISTANCES FROM THE RIM OF THE SUN<br>4 scales, in metres and centimetres | | | |
|---------|---------|---------|---------|---------|
|         | Scale 1 | Scale 2 | Scale 3 | Scale 4 |
| MERCURY | 20 cm | 40 cm | 60 cm | 80 cm |
| VENUS | 38 cm | 76 cm | 1 m 14 cm | 1 m 52 cm |
| EARTH | 50 cm | 1 metre | 1 m 50 cm | 2 metres |
| MARS | 80 cm | 1 metre 60 cm | 2 m 40 cm | 3 m 20 cm |

4   Collect for notebooks or general display, up-to-date information and illustrations about artificial satellites.

**Written Work**

1   A planet goes round a star.
2   A satellite goes round a planet.
3   The sun has nine planets going round it.
4   Mars has two satellites. Mercury and Venus have none.
5   Planets and satellites shine in the sun's light.
6   A star is made of glowing gases.

# DEAD THINGS ARE USED BY LIVING THINGS

## THE MAIN USES FOR DEAD THINGS

### Demonstration Material

Any or all of the following:

1   Fungus feeding on dead material, *e.g.* mould or mildew on old leather, fungus on a twig or piece of tree bark.
2   Material which has been partly used for food by an animal, *e.g.* wool or felt containing holes made by the larvae of a clothes moth, or piece of wood containing holes made by the larvae of a timber beetle.
3   Any example of a protective shelter built by an animal, *e.g.* a hamster's nest, a bird's *old* nest, or the larva of a caddis fly in a case made out of dead plant material.

### Sample Link Questions

1   What are the three main kinds of things? (*Alive, dead, never alive*)
2   What are the two kinds of dead things? (*Dead animal, dead plant*)
3   Are fur, feathers and leather dead animal or dead plant parts? (*Dead animal parts*)
4   Are tea leaves, cotton and wood dead animal or dead plant parts? (*Dead plant parts*)
5   Are cooked foods alive or dead? (*Dead*)
6   What two things do living animals and plants need in order to grow? (*Oxygen and food*)
7   What is the name of a moth whose larva feeds on wool? (*Clothes moth*)
8   What are the two main reasons for animals carrying things from place to place? (*They carry things for food, and they carry things to build with*)
9   Is a fungus a plant with roots, stems and leaves, or is it a simple plant? (*A simple plant*)
10  What do many birds build to protect their young? (*Nests*)

## Relevant Information

In Pupils' Book 1 a lesson on dead material established that it is of two kinds – dead animal and dead plant.

The main points of this lesson are:

1   dead material is used mainly for food by certain kinds of plants and many kinds of animals – including human beings
2   dead material is used for protection by certain animals – including human beings
3   dead material is used in additional ways by human beings alone.

The main ways in which human beings use dead things is the subject of a lesson in Pupils' Book 3. The methods by which we protect dead material from other living things in order to preserve it for our own use, are the subject of a separate lesson in Pupils' Book 4.

When a living plant or animal ceases to respire (*i.e.* to use oxygen) it dies. Once an organism is dead, its decomposition begins. The decomposition may be assisted by desiccation (*i.e.* evaporation of water) or chemical action (*e.g.* oxidation), but it is brought about mainly by the activities of living things. If it were not for the uses to which living things put dead material, the world would probably be littered with the remains of dead animals and plants.

### Dead Things are Used Mainly for Food

Most of the living things which use dead material use it for food. Members of both the plant kingdom and the animal kingdom are responsible.

### Plants which feed on dead things

Green plants – those containing chlorophyll – obtain their food from such never-alive materials as water, dissolved salts and the carbon from carbon dioxide. These include the plants with roots, stems and leaves, together with two classes of simple plants – the algae, and the moss and liverwort class.

There are, however, two very important classes of simple plants which have no chlorophyll and which rely for their food on dead things or on living things. They are the bacteria and the fungi.

*a Bacteria.* Bacteria are not mentioned in this lesson in the Pupils'

Book. They are one-celled plants, and are amongst the smallest of living things. They are so small that an average full-stop in a child's notebook would be large enough to cover over a thousand of them.

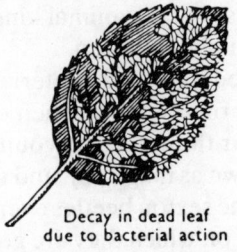

Decay in dead leaf
due to bacterial action

A can of food spoiled by bacteria. The resulting gases have caused the can to bulge

Although too small to be seen individually, developing colonies of them may be large enough to be seen by the naked eye, as may the decay resulting from their activities, *e.g.* on a dead leaf being reduced to a skeleton.

*b Fungi.* Fungi include mushrooms, toadstools, moulds, mildews, rusts, blights and yeasts. They are widely distributed throughout the world. Those which feed on either living plants or living animals are termed parasites. Those which feed on the remains of dead

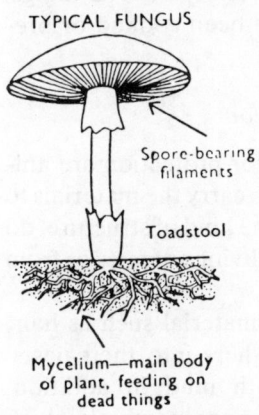

TYPICAL FUNGUS

Spore-bearing filaments

Toadstool

Mycelium—main body of plant, feeding on dead things

plants or dead animals are termed *saprophytes*. Dry rot, which causes so much damage to timber, is an example of a fungus feeding on dead material. The main body of a fungus consists of tubular threads which penetrate the tissues of the host. These tubes, known as hyphae, form a network into which food is absorbed. When mature, the fungus produces specialised hyphae which grow away from the food and terminate in swellings containing the reproductive spores. These spores are carried by air, water, or animals to other sources of food.

Bacteria and fungi are also found on manufactured substances containing dead material, *e.g.* bread, jam.

The obvious signs of decomposition brought about by bacteria or fungi are often referred to as rot or decay. In the case of bacteria especially, the breaking down of tissues may result in the release of gases with an obnoxious smell.

275

## Animals Which Feed on Dead Things

No animal is able to feed like the green plants; all animals must rely for food on living or dead material. Those which feed on dead material are to be found amongst all classes of the animal kingdom, ranging from the simplest to the most highly evolved.

Among the small animals which feed on dead materials are certain insects, mites, crustaceans and worms. Some lay their eggs in dead material to ensure a food supply for the emerging young, *e.g.* the clothes moth, bluebottle (larvae known as maggots), and timber beetle (larvae known as woodworms). The sexton beetles even go to the trouble of burying the small carcases on which they are going to lay their eggs. In the temperate and warmer regions of the world ants dominate among the insect scavengers. They patrol almost every square metre of ground in which plants grow and carry home for food well over 80% of the dead insects which can be found there.

Some animals which prey on other living animals kill their prey immediately before they consume it, but they cannot strictly be included amongst those which feed on dead things. On the other hand, birds like the vulture and mammals like the jackal feed on other animals which are already dead.

Human beings, of course, feed on a wide variety of dead things. The parts of animals and plants which have been cooked or preserved are obvious examples of this.

## Some Animals Use Dead Things for Protection

The only living things to collect dead parts for protection are animals. It is only they, of course, who are able to carry the materials to build with. Those animals which do make some kind of structure, do so in order to provide protection, either from living enemies or from the weather or, in some cases, from both.

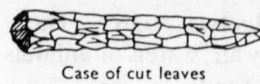

Case of cut leaves

Birds incorporate material such as hair, straw, twigs and feathers into their nests. Some line them with mud in addition. Squirrels and hamsters are mammals which build themselves nests from dead material. A bird's nest serves not only to protect the eggs but also to protect the helpless young from enemies. Caddis-fly larvae construct their cases out of various materials found in the water which they occupy. These cases,

Case of cut stems

TYPICAL CASES OF
CADDIS FLY LARVAE

which serve as camouflage, are in many instances constructed from strips of dead plant material such as bits of leaves and twigs.

Human beings are outstanding examples of animals which use dead things for protection. Wood in particular has been used throughout human history for the building of protective fences and stockades, as a defence against avaricious human beings and wild animals. The Normans, with their early towers and stockades of wood, and the U.S. fifth cavalry, with their timber forts, were using dead material for protection against enemies.

Amongst modern communities the main use of wood for protection is against the weather. The skins of various mammals are also used in different parts of the world for this purpose. The tupik or summer dwelling of Eskimos, for example, is a tent consisting of seal skins sewn together, and stretched over a framework of poles. The tepees of North American plains Indians are also made out of skins. (A wigwam, strictly speaking, consists of a framework of poles covered with bark.) Nomads in the desert use cloth for their tents, and the cloth itself is made from the hair of mammals, particularly the hair from camels together with the wool from sheep. Nomads in Soviet middle Asia construct a tent known as a yurt. It consists basically of a light wooden framework which can be folded and carried. It is covered with a matting of woven reeds in summer and with felt in winter. Felt itself consists of matted fibrous material made up chiefly from mammal hairs of the kind which form a thick fur. In tropical countries, such as Pakistan, the stems of coarse grasses are matted together to form materials for building huts. The dead stems of coarse grasses are also used for thatching. Even in the British Isles thatched roofs are still in evidence.

Clothing, much of which is made from dead animal or dead plant parts, also gives humans protection against the weather.

In the animal kingdom in general, however, far fewer species use dead things for protection than use them for food.

## Only Humans Use Dead Things in Other Ways

Human beings are the only living things to use dead material for purposes other than food or protection. In addition to food and protection, perhaps the two most important uses to which dead materials have been put throughout human history are fuel and vehicles for transport. Clothing, which is really a form of protection, is of course also peculiar to humans.

The illustrations for Lesson 31 in the Pupils' Book show:

1 Living plants feeding on dead plant material – fungi on part of a tree trunk and on dead twigs:
(*Top*) a common gill fungus (*Hypholoma hydrophilum*)
(*Left*) a common bracket fungus (*Polystictus versicolor*)
(*Right*) the dryad's saddle fungus (*Polyporus squamosus*)
(*Bottom – on twigs*) bird's nest fungus (*Crucibulum vulgarae*).

2 Living animals feeding on dead plant material – the larvae of the common furniture beetle (*Anobium punctata*) burrowing through wood.

3 Living plants feeding on dead animal material – common mould on a leather belt.

4 Living animals feeding on dead animal material – clothes moth larvae on material made out of wool.

5 Dead animal and plant material (cooked) used as food by human beings: boiled eggs, baked beans and fried bacon.

6 Dead plant material used for protection against enemies – a blackbird's nest made from grass and twigs.

7 Dead plant material used for protection against enemies – a caddis-fly larva in a case made out of bits of freshwater herbs.

8 Dead plant material (log cabin), and dead animal material (in the form of fur and wool) used as protection against the weather.

## CODE

1 Observe any examples of dead material being used as food by a living plant, *e.g.* dead plant or animal parts being fed upon by fungus.

2 Observe any examples of dead material being used as food by a living animal, *e.g.* wool or felt containing small holes made by feeding moth larvae, piece of wood containing small holes made by the larvae of timber beetles, the different kinds of animal and plant parts which we cook and feed upon.

3 Collect for a frieze, labels from packets and tins of different kinds of animal parts and plant parts upon which we feed.

4 Observe the kinds of dead materials incorporated in a bird's nest or a pet hamster's nest.

5 Collect any illustrations which show dead animal or dead plant material being used as protection against enemies or the weather.

**Written Work**

1 Dead things are mainly used for <u>food</u>. Some <u>simple</u> plants and many <u>animals</u> feed on them.
2 Some <u>animals</u> use dead material to protect themselves from <u>enemies</u> or from the <u>weather</u>.
3 Only <u>humans</u> use dead material in other ways.

# 32  ROCK, WIND AND WATER

## ROCK IS WORN DOWN

**Demonstration Material**

1 Sands of various grades, *e.g.* fine and medium.
2 Aquarium gravel.
3 Pebbles and stones.
4 A magnifying glass, if possible.

**Sample Link Questions**

1 Of what are pebbles and stones small pieces? (*Rock*)
2 What do we call big pieces of rock? (*Boulders*)
3 What name do we give to tiny grains of rock? (*Sand*)
4 What do we call the rock which is made of tiny grains of sand stuck together? (*Sandstone*)
5 Are most rocks a mixture of solids, or of liquids, or of gases? (*Solids*)
6 What is mixed with certain kinds of rock powder to form the paste we call clay? (*Water*)
7 What are the three weather makers? (*Sun, air, water*)
8 Moving animals carry things from place to place. What are the two other things which do this too? (*Moving air, moving water*)
9 What forces things towards the earth? (*Gravity*)
10 What are streams and rivers flowing towards? (*The sea*)

**Relevant Information**

A lesson in Pupils' Book 1 emphasised that boulders, stones, pebbles and grains of sand are all pieces of rock. The main points of this lesson are:

1  that moving air and moving water are the two main agents responsible for wearing down rock and pieces of rock into boulders, stones, pebbles and sand;
2  that moving air and moving water are also the two main agents responsible for conveying them from place to place.

The thin crust of our planet consists of solid rock. Most rocks are a mixture of two or more minerals, although some rocks consist almost entirely of one mineral.

For nearly three-quarters of the surface of our world this thin crust of rock is covered by the waters of the sea. The fraction which is raised above the level of the sea, and which we call land, may itself have areas covered by water in the form of ice, snow, lakes or rivers. In some places it is covered by sand, which has worn down from rock, and in others it is covered by a thin veneer of soil, the basic ingredients of which are themselves fine particles of rock.

The face of the land is forever changing. The process is very slow, relative to the span of a human life, but it is taking place gradually all the time. This change is mainly the result of two opposing processes.

A  The crust is gradually being folded and raised up, owing to various actions below it.
B  The crust is gradually being broken down and levelled, owing to various actions above it.

A number of factors contribute to the breaking down or denudation of the crust. These include:

1  The erosion of the rock due to:
   a disintegration and splitting resulting from such factors as:
      i changes in temperature. This is particularly noticeable in high mountainous regions, where temperature changes are severe, as a result of frost and the heat of the sun.
      ii chemical action brought about mainly by such gases in the air as oxygen, water vapour and carbon dioxide.
      iii the pressure exerted by plant roots.
   b general wearing down by moving air and moving water.
2  The moving of particles of rock due to the forces of:
   a gravity
   b moving air and moving water.

As gravity is responsible for the movement of both water and air, it is, of course, the dominant factor.

The expressions *boulder, stone* and *pebble* are all used to describe detached fragments of rock, but there is no strict definition in terms of size or shape. The term *boulder* is generally applied to a large weathered fragment of rock with a diameter of more than 25 centimetres. The term *stone* is generally applied to a smaller fragment and the term *pebble* to a small stone worn smooth by the action of water. The word *stone* is particularly applied in connection with rock used for building purposes, sections of rock cut to a specific size or shape (*e.g.* gravestone, paving stone) and pieces of rock composed of minerals which, for colour, beauty or rarity, have an ornamental value (*e.g.* precious stones and gem stones).

The terms *sand, gravel* and *shingle* are applied to large quantities of small stones or pebbles, and again there is no strict definition. Generally speaking, a collection of tiny fragments is called sand, and a collection of coarser fragments is called gravel or shingle.

## The Wearing Down of Rock on the Land

The wind and rain play a large part in the weathering down of inland rocks. In dry regions dust and other fine particles of detached rock are carried by the wind and deposited in large quantities, often great distances away. This carrying of rock particles by moving air can be seen and felt among sandhills on a windy day. Sandhills are common on coasts where the prevailing wind is an onshore one, blowing the sand of the beach inland.

Great sandy wastes like the Sahara desert and the Gobi desert have been formed by the carrying and depositing of particles of worn down rock, and it is in the desert that the wind is the main agent of transport.

Rock particles which are carried by the wind act as an abrasive in wearing down other rocks and, in arid regions especially, the wind plays a large part in wearing down rocks by driving sand against them.

Where an exposed formation of rock is a mixture of hard and soft rocks, the softer rocks are worn down first and the harder sections more slowly. Odd formations of rock often result from this more rapid wearing down of the softer layers in an exposed section. Rock arches and the desert buttes seen in cowboy films are examples.

HOW SHAPES ARE CARVED OUT OF ROCK

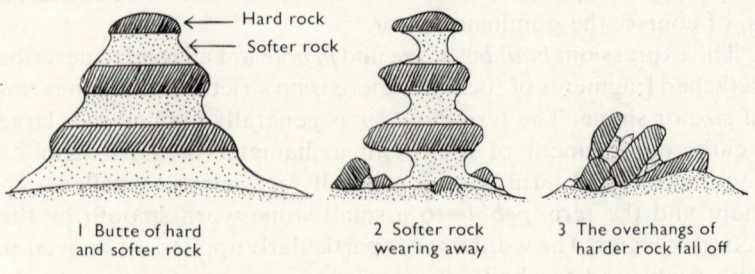

1 Butte of hard and softer rock

2 Softer rock wearing away

3 The overhangs of harder rock fall off

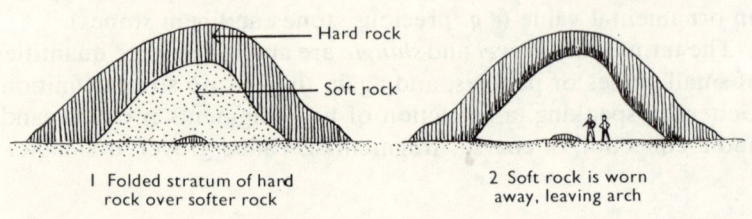

1 Folded stratum of hard rock over softer rock

2 Soft rock is worn away, leaving arch

## The Wearing Down of Rock in Streams and Rivers

Water is largely responsible for transporting loose rock material towards the sea. Rain water washes it down the slopes into brooks and streams. Brooks and streams carry it into rivers, and these in turn convey it towards the sea. Streams and rivers carry solid matter in three ways:

1   in solution, *e.g.* salt
2   in suspension, *e.g.* sand, and even large particles whose weight is insufficient to withstand the moving water
3   by dragging larger fragments along its bed.

The particles which are carried in suspension assist in wearing away the sides of the stream or river, and the larger fragments which are dragged along the bottom assist in wearing away the bed, so that the course along which the water flows is worn wider and deeper. The Grand Canyon has been carved out by the Colorado River in this way during about a million years. In places it is 20 kilometres wide at the rim and over 2 kilometres deep.

Where a waterfall occurs, considerable erosion may result. This often takes place in two stages:

1   the falling water wears away the base and undermines the cliff over which it falls;
2   the upper part, lacking support, collapses.

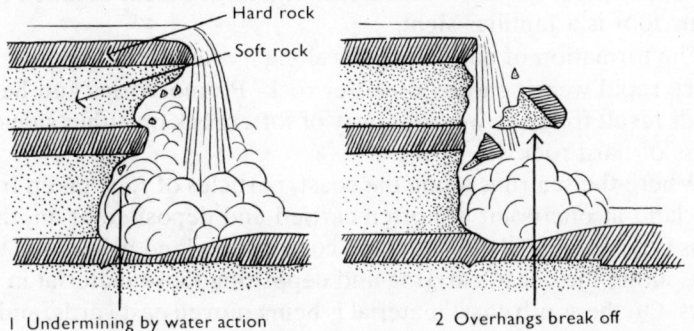

1 Undermining by water action 2 Overhangs break off

By this method a waterfall is constantly wearing its way backwards towards the source of the river. Niagara Falls, for example, are estimated to be wearing their way backwards at the rate of about one metre a year.

In the Arctic and Antarctic, or in high mountains where glaciers are formed, water remains the main agent of transport and erosion, although it is in the solid form. Like a river, a glacier affects rock in three main ways:

1   it carries fragments of eroded material towards the sea;
2   it erodes the surface with which it is in contact. This is done mainly by the rock fragments dragged at the bottom of the glacier;
3   it deposits some of the carried fragments along its course; and where its tip finally melts, it leaves a mass of debris known as a moraine.

V-shaped valleys are worn into U-shaped valleys by glacial action. The fiords of Norway are examples of glacial action in the past.

### The Wearing Down of Rock by the Sea

The surface of cliffs along the coast are of course eroded by the same agents as an inland surface, but where the sea washes against the foot of the cliffs, then breaking waves are chiefly responsible. The waves themselves are caused by the wind blowing across the surface of the sea.

Where the sea reaches the foot of a cliff, it eventually wears away the foot of the cliff, assisted by the rock particles which it carries. Such constant undermining subsequently results in the collapse of the upper part. The sea-washed cliff with large blocks of fallen rock at its foot is a familiar sight.

The formation of bays and gulfs along a coast is the result of the more rapid wearing away of softer rock. Promontories and headlands result from the wearing away of softer rock on either side of a mass of hard rock.

Where the sea runs along the coast, particles of rock taken from the land at one point are often carried and deposited at another. This is noticeable along the south coast of England, where the tide runs along the coast, carrying and depositing loose material in the bays. On the south coast material is being moved eastwards, and on the east coast material is being moved southwards in this way.

The actual rock on the shore itself is constantly being worn down by breaking waves, so that boulders, pebbles and sands are all in the process of being ground down into smaller pieces. The sea is assisted in this work by the rock particles which it carries. When lashed by a storm, the sea can have tremendous power, and it has been known to hurl boulders weighing up to 40 tonnes.

## CODE

1    Observe stages in the wearing down of small pebbles into sand, by examining a number of different grades of gravel and sand. These may be kept in separate specimen tubes or small jars for observation purposes.

2    Collect different shapes of pebbles, as well as different colours. Shapes may be ovoid, flat, spherical, or just odd.

*Note*

The composition of a pebble sometimes influences the shape into which it is worn. Slate, for example, is usually flat as a pebble. The ovoid shape is fairly common, and the spherical is least common.

3    Observe any broken pebbles. After being constantly ground against other pebbles, the outer surface of a pebble develops a deceptive 'skin' which disguises its true nature. Newly-broken pebbles are more revealing. A pebble may be broken for examination by wrapping it in a piece of cloth, and striking it with a hammer. The cloth wrapping prevents flying particles.

4   Experiment to show, by rubbing, that some rocks are softer than others, and therefore wear down more easily, *e.g.* clay, chalk and shale are softer than flint and granite. Comparisons may be made by rubbing with fingers, and by rubbing with the sandpaper strip on a matchbox.

5   Collect samples of sand and gravel from different shores and beaches. Keep each sample in a separate specimen tube suitably labelled with the place of origin and the date.

*Notes*

*a  i* for fine sand, 75 mm × 19 mm specimen tubes are large enough

  *ii* for coarse sand and fine gravel, 75 mm × 25 mm tubes are suitable

  *iii* for casual beach profiles, use taller tubes, *e.g.* 100 mm × 25 mm, 125 mm × 25 mm, and 150 mm × 25 mm – or even large Alka Seltzer tubes.

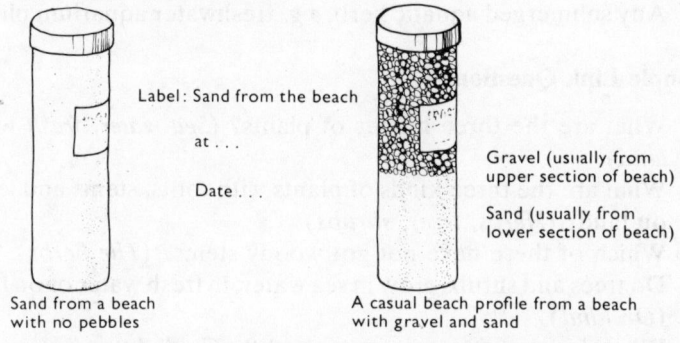

Label: Sand from the beach

at . . .

Date . . .

Gravel (usually from upper section of beach)

Sand (usually from lower section of beach)

Sand from a beach with no pebbles

A casual beach profile from a beach with gravel and sand

*b* In a quantity of sand are grains of various minerals, but quartz is usually by far the commonest. There are several reasons for this: quartz is one of the most abundant minerals in rocks; it is comparatively hard; it is only slightly soluble in water and therefore will remain after the grains of certain other minerals have dissolved.

**Written Work**

1   Moving <u>air</u> and moving <u>water</u> wear down rocks. They <u>carry</u> the pieces from place to place.

2   Desert sands are carried by the <u>wind</u>. The sands of the shore have been worn by <u>water</u>.

3   Wind <u>forces</u> the water into waves.

# 33 LIVING PLANTS IN FRESH WATER

## HERBS AND SIMPLE PLANTS

### Demonstration Material

Any or all of the following:

1 Freshwater algae, *e.g.* pond 'slime', or 'blanket weed'.
2 Aquatic moss, *e.g.* great water moss (*Fontinalis*).
3 Fungus growing on any dead material left in fresh water.
4 Swamp herb.
5 A freshwater herb or part of a herb with floating leaves, *e.g.* water buttercup, water lily leaf and stalk.
6 Floating herb, *e.g.* duckweed, frogbit.
7 Any submerged aquatic herb, *e.g.* freshwater aquarium plants.

### Sample Link Questions

1 What are the three homes of plants? (*Sea water, fresh water, land*)
2 What are the three kinds of plants with roots, stems and leaves on land? (*Herbs, trees, shrubs*)
3 Which of these have not got woody stems? (*The herbs*)
4 Do trees and shrubs grow in sea water, in fresh water or on land? (*On land*)
5 What do most stems grow towards? (*The light*)
6 Do roots grow towards the light or away from it? (*Away from it*)
7 What do we call plants which have no roots, stems or leaves? (*Simple plants*)
8 Name a simple plant which feeds on living things or on dead things. (*Fungus*)
9 What are the four main needs of living things? (*Oxygen, food, to grow, to have young*)
10 What is the difference between sea water and fresh water? (*Sea water has more salt in it*)

### Relevant Information

The purpose of a lesson in Pupils' Book 1 was to show that the three homes of plants are

*a* sea water,
*b* land, and
*c* fresh water.

Lesson 3 in this book dealt with the different kinds of plants found on land, namely herbs, trees, shrubs, and simple plants. The main purpose of this lesson is to show that

1 both herbs and simple plants are found in fresh water
2 the aquatic herbs are grouped according to habit:
   *a* swamp herbs
   *b* rooted herbs with floating leaves
   *c* floating herbs
   *d* underwater herbs

### The Herbs

Land plants do not move from place to place to seek their food. There is, in general, no need for them to do so. They depend upon the rain, and are in consequence at a certain disadvantage in the event of a drought. Life in or near water serves as an insurance against shortage of water. As roots, stems and leaves were a development of the land plants, those herbs which are found growing in fresh water are looked upon as being land plants which have adapted themselves to these conditions. Some aquatic herbs, such as flote grass, can readapt themselves in dry spells to terrestrial conditions, although growth may be stunted.

In any area of water various factors are responsible for determining the extent to which aquatic herbs are found. A sufficiency of oxygen, food and light are necessary. In stagnant pools, where there is decaying vegetation from overhanging trees, oxygen may be in short supply owing to the presence of bacteria. Some water may be too acid, or too alkaline; it may be too hard or too soft; or it may be too cloudy for sunlight to penetrate satisfactorily. It may be moving swiftly, in which case only certain well-anchored plants may be established there.

Most aquatic herbs are perennials, and they have various ways of surviving the winter. Certain plants, for example, water starwort and water soldier, sink to the warmer water at the bottom of the pond in autumn, where they remain until spring. Others, such as frogbit, water milfoil and bladderwort, develop tightly packed buds in autumn. When the parent plant decays, these buds sink to the bottom, where they remain for the winter. During the following spring, they rise to the surface and develop into new plants. Many other plants store food in swollen stems which are buried in the mud. Water lilies are examples.

Certain land plants known as marsh plants grow where firmer land and swamp merge. These are not true aquatic herbs, but their roots are in soil which is nearly always moist. Trees such as alders and willows are often found in these regions. Such plants are *hygrophytes*, and should be distinguished from the true aquatic herbs, which are *hydrophytes*.

Trees and shrubs are not generally found amongst the plants which have adapted themselves to freshwater conditions, and this is especially true in the British Isles. The bald cypresses and the mangroves of the swamps of Florida are exceptional, and even they have the greater part of their woody stems above water level.

## Swamp Herbs

These are the least adapted to fresh water. They grow in the shallows with their roots in the mud. The swampy area is usually covered with water, except during the driest parts of the year. Swamp plants are generally tall, with creeping underground stems which help to keep the main part of the plant upright.

The swamp herbs illustrated in this lesson in the Pupils' Book are:

a *Yellow Iris* – also known as the yellow flag (*Iris pseudacorus*). This is the commonest wild iris. Its stems are stiff and erect. The leaves which are sword-shaped, are also stiff and erect. The flowers are large, and a bright yellow in colour, and are found in June to August.

b *Arrowhead (Sagittaria sagittifolia)*. This plant gets its name from its aerial leaves which stand above the water on erect stalks. There are also oval-shaped floating leaves and ribbon-like submerged leaves. The flowers are borne in threes at the top of an erect flower stalk.

## Rooted Herbs with Floating Leaves

These plants, together with the underwater herbs and the floating herbs, lack the tough fibres which are necessary to enable the stem of a land plant to stand upright. Tiny pores on the upper surface of

## SOME FRESHWATER PLANTS

**REED-MACE**
Brown spike of
female flowers.
Male spike
above it

**COMMON REED**
Purple flowers

**BUR-REED**
Leaves keeled and
triangular at base.
Green flowers

**SWEET FLAG**
Spike of green
flowers. Thick
stem

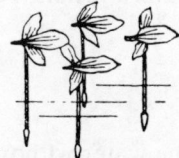
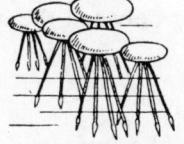
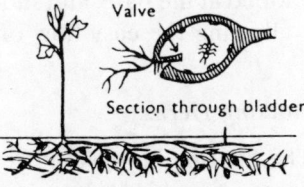

Valve

Section through bladder

**IVY-LEAVED DUCKWEED**
Just under surface.
One root per plant

**GREATER DUCKWEED**
Floats on surface.
Several roots. Purple
underside

**BLADDERWORT**
Submerged leaves bear
bladders. Small insect
larvae sucked in through
valve. Yellow flowers

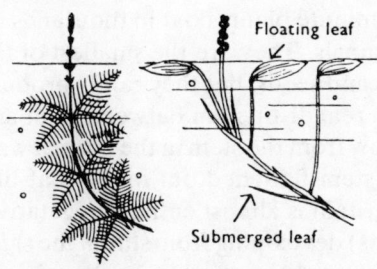

Floating leaf

Submerged leaf

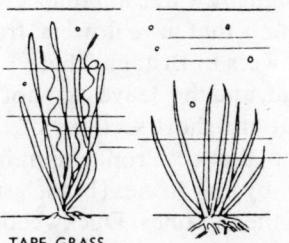

**WATER MILFOIL**
Submerged herb.
Green flower-
spike projects
above water

**BROAD-LEAVED
PONDWEED**
Two kinds of leaf.
Spike of green
flowers

**TAPE GRASS**
Spiral stalks
of female
flowers shown.
Pale pink male
flowers break
off and float

**QUILLWORT**
Spore-bearing
herb with spores
in capsules on
leaf bases

*Row 1—*Swamp plants    *Row 2—*Floating plants    *Row 3—*Submerged plants

*Note.* The insect sucked into the bladderwort dies and decomposes.
The walls of the bladder then absorb substances from its body.

floating leaves enable them to obtain their oxygen from the air.

The two herbs of this kind that are illustrated in the Pupils' Book are:

a *Yellow Water Lily (Nuphar lutea)*. This, together with the white water lily, is commonly found growing wild in Britain. Neither is a true lily. The large circular leaves are borne on long leaf stalks which are generally at an angle to allow for the rise and fall of the water. Oxygen is obtained from the air through tiny pores in the upper surface. The white water lily produces the largest British wild flowers. Seeds are carried away by moving water. The plant dies down in winter, with food stored in a swollen stem buried in the mud. Further propagation takes place from this stem.

b *Water Crowfoot (Ranunculus)*. These are sometimes called water buttercups, and the flowers appear from May to August. The leaves are of two kinds – floating, kidney-shaped leaves that are lobed at the base, and submerged leaves that are thin and hairlike, allowing the easy flow of water between them.

## Floating Herbs

The floating herbs float freely on the surface of the water without being rooted in the mud at all. The roots hang loose, take in food and water, and also serve as a means of maintaining balance.

Those illustrated in this lesson in the Pupils' Book are:

a *Duckweeds (Lemnae)*. These minute plants float in thousands on stagnant water in ponds and canals. They are the smallest of the plants that have flowers, fruits and seeds, but they rarely produce flowers in Britain. There is no real distinction between stem and leaf, and the 'leaves' do not grow from the stem in the normal way, but are short sections of the stem flattened out into a leaf-like appearance. Propagation in Britain is almost entirely vegetative, *i.e.* by new fronds (branch stems) developing from slits in the sides of the old ones. Duckweeds obtain their oxygen from the air and may thrive on the surface of waters which are too deficient in oxygen to support other herbs.

b *Frogbit (Hydrocharis morsus-ranae)*. The small circular leaves grow in the form of a rosette. During summer new plants grow at the ends of branch stems, but during autumn buds form at the ends of these branch stems instead. These become detached and sink for the winter, rising in the following spring to form new plants.

The white flowers are in evidence in July and August.

c *Water Soldier (Stratiotes aloides)*. This plant gets its name from the sharp sword-like leaves which rise from a short central stem. For most of the year the plant floats below the surface, but about July it rises, so that its leaves and white flowers are well above the surface. Propagation in Britain is by means of new plants developing from branch stems. Within the British Isles the water soldier is found mainly in eastern England.

*Note*

The rising and sinking of plants in the water is, of course, not due to self-propulsion. For example, frogbit buds at the bottom of a pond become lighter when some of their food reserves are used up, and they rise to the surface in spring as a result. The movement of the water soldier is a little more complicated. It grows in hard water containing calcium carbonate and, as a result of photosynthesis, a deposit of this forms on the leaves, the extra weight causing the plant to sink. In spring, when new leaves are growing, they eventually become more numerous than the old ones and, not having the same proportion of deposited solids, increase the buoyancy of the plant, eventually causing it to rise to the surface again.

*Underwater Herbs*

The submerged herbs are the most completely adapted to aquatic conditions, and various modifications to roots, stems and leaves may have taken place. It is usual for oxygen as well as water and dissolved foods to diffuse into the plant tissues through the surface of the stems and leaves. In consequence, air holes are not required on leaves, and the function of roots may be chiefly that of anchorage. Although flowers may continue to be produced, reproduction often takes place chiefly as a result of new plants growing from part of the parent. It is the submerged herbs which are used mainly for decorative effects in cold and tropical aquaria.

Those illustrated in this lesson in the Pupils' Book are:

a *Canadian Pondweed* or *Water Thyme (Elodea Canadensis* – once termed *anacharis)*. This plant, introduced from North America, reproduces readily by means of new plants developing from the stems of the old. On the many branch stems the leaves are small, simple and lance-shaped, and usually in whorls of three. In Britain flowers rarely appear.

291

b *Starwort (Callitriche)*. The leaves are bright green. The lower leaves are long and narrow, but those which reach the surface become broader and flatter, forming a star-shaped rosette from which the plants get their name. The flowers are small and insignificant.

c *Hornwort (Ceratophyllum demersum)*. This herb is so completely adapted to fresh water that it is able to absorb water and dissolved foods directly through its surface, so that true roots no longer appear, although lightish shoots which penetrate the mud for anchorage purposes are sometimes developed. The plant usually floats in an erect position, however, without being anchored at all. The finely divided leaves grow from the stem in whorls, and old leaves appear horny – hence the plant's name. The inconspicuous flowers, which are developed under water, seldom appear on plants in Britain. New plants develop readily from part of the parent.

*Note*

Some underwater herbs are occasionally sold by dealers as 'oxygenating plants', the possible implication being that this is a special characteristic of certain varieties. It is a fact of course that *any* plant containing chlorophyll releases oxygen during the feeding process of photosynthesis. Some of this oxygen is used by the plant during respiration, and any surplus becomes 'free'. Such plants include all the herbs, trees and shrubs, together with those simple plants which are algae or which belong to the moss and liverwort group. The plants which do not contain chlorophyll are the fungi and the bacteria.

**The Simple Plants**

Representatives of three of the main groups of plants without roots, stems and leaves can be seen in fresh water namely:

1 the algae
2 the mosses and liverworts
3 the fungi.

The fourth main group of simple plants, the bacteria, are also represented in fresh water but, because of their smallness in size, are not included in the Pupils' Book. Evidence of their presence is sometimes apparent, however, especially in water which is cloudy or smelly and in a state which we term *foul*. Both bacteria and fungi need oxygen like any other living thing, and where there is a considerable amount of decay due to their activities, the surrounding water may be lacking in oxygen as a result, and thus be unhealthy for other living animals and plants. Such circumstances may be brought about by over-feeding fish in an aquarium.

## Algae

These are numerous in fresh water. They have chlorophyll like the herbs, and their food requirements are therefore similar to those of the herbs. They appear unsightly in an aquarium, but in general they are beneficial to animal occupants. Many are floating, but some kinds grow anchored to stones or even to the glass sides of a container. Both types are illustrated in the Pupils' Book.

## Mosses and Liverworts

These are the highest of the simple plants, and may show symptoms of leaf and stem-like growth. Few species actually live in fresh water, the best-known of these being the sphagnum or bog mosses, which float, and the willow mosses, which grow anchored and submerged.

The illustration in the Pupils' Book shows water moss (*Fontinalis*), sometimes called willow moss. This plant forms dark green trailing masses, often attached to stones. Reproduction can be by spores but is mainly by means of new plants growing from part of the parent.

## Fungi

Fungi are sometimes called water moulds. Some – the saprophytes – are beneficial as they feed on and break down dead animal or plant tissue, but others are parasitic on other living plants or animals.

The illustration in the Pupils' Book shows saprolegnia, the fish fungus which attacks living fish, especially where there has been damage to scales or a slight wound. Goldfish and other fish are subject to the attacks of this fluffy fungus. It is usually killed by immersing the fish in a solution of salt and water calculated to be strong enough to kill the plant without killing the fish.

293

## CODE

1   Collect freshwater plants into sets, *e.g.*
   *a* freshwater herbs
     *i* floating herbs
     *ii* underwater herbs
   *b* freshwater simple plants
     *i* algae
     *ii* willow moss
     *iii* fungi found growing below the surface.

*Notes*

*a* Rooted herbs with floating leaves and swamp herbs are generally too large to be kept conveniently in a classroom for any length of time.

*b* Other freshwater plants are probably the easiest plants to maintain in the classroom, as they require little or no attention.

*c* Large jars, *e.g.* toffee jars, are suitable containers for freshwater herbs, and a layer of aquarium gravel is usually all that is necessary for rooting purposes. A pebble or a small piece of lead may be used to hold the more awkward ones in position. Such miniature aquaria make useful temporary homes for specimens of aquatic animal life. For observation purposes, these aquaria are best kept in indirect light. If they are exposed to direct sunlight, the possible formation of algae will tend to spoil their appearance.

2   Observe the different ways in which freshwater herbs reproduce from part of the parent, *e.g.* from one duckweed, one frogbit, one section of Canadian pondweed stem.

3   Observe that willow moss (*Fontinalis*) floats submerged in water. Observe new plants growing from part of the parent.

4   Observe development of algae in jars of water – especially pond water – kept in strong sunlight. Containers with algae may be kept indefinitely.

5   Observe development of fungus on dead material left in water. The smell caused by bacteria may also be observed.

6   Preserve specimens of aquatic plants in 5% formaldehyde. (See 'Keeping Specimens for Observation Purposes' page 27 for information on use of copper acetate.) A small fish which has died as a result of fungus attack may be preserved in 5% formaldehyde – for observation of the fungus.

**Written Work**

1 Freshwater plants are either <u>herbs</u> or <u>simple</u> <u>plants</u>.
2 <u>Fungi</u> feed on living or dead animals and <u>plants</u>.
3 The <u>stems</u> and <u>leaves</u> of swamp herbs grow out of the water.
4 Floating leaves take oxygen from <u>the air</u>.
5 Underwater herbs use dissolved <u>oxygen</u> and <u>food</u>.

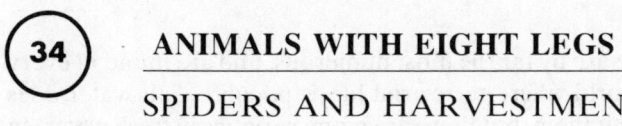

## 34  ANIMALS WITH EIGHT LEGS

### SPIDERS AND HARVESTMEN

**Demonstration Material**

Any living or preserved spiders and harvestmen.

**Sample Link Questions**

1 What are the three regions where living animals can move from place to place? (*On land, in water, in the air*)
2 Which class of animals drink milk when they are young? (*Mammals*)
3 Which class of animals has feathers? (*Birds*)
4 To which class belong the animals which have both fins and gills? (*Fish*)
5 To which class of animals do frogs, toads and newts belong? (*Amphibians*)
6 Which class of animals have scaly skins and lungs? (*Reptiles*)
7 Which class of animals have six legs when they are adults? (*Insects*)
8 Which insects have powdery scales on their wings? (*Moths and butterflies*)
9 To which family of insects do the bluebottle and crane fly belong? (*The two-winged flies*)
10 How many feelers has an insect? (*Two*)

## Relevant Information

The main purpose of this lesson is to introduce the class of animals with eight legs, exemplified by the spiders and harvestmen.

### Arthropods in General

These animals with their jointed limbs and armoured bodies, were the first of the land animals. They evolved from the ringed-worms, and like these worms, still have a body divided into segments. 'Arthropod' – meaning *jointed limbed* – is an unfortunate name, as other animals outside this group have jointed limbs also. There are four classes:

*Insects*. These are by far the most numerous, and are found in every part of the world where terrestrial life is possible. Salt water does not seem to suit them, but there are many varieties in fresh water. In the adult stage, they all have six legs and two feelers. Most have four wings, some have two wings, and a few species have no wings.

*Arachnida*. These arthropods have eight legs. The class includes spiders and harvestmen (harvesters), false spiders, scorpions, false scorpions, and mites. No arachnid has wings. Most of them are land animals, but some are found in fresh water, and others are found in the sea.

*Crustaceans* (Crusty skins). The number of legs varies, but there are never wings. The class includes crabs, lobsters, shrimps, and water lice. They are found mainly in sea water, but some live in fresh water and certain species such as the woodlouse are terrestrial.

*Myriapoda*. These are the arthropods with many legs, *i.e.* the centipedes and millepedes, and they bear certain similarities to their worm-like ancestors. None has wings, and they are all land animals.

*The Arachnida*. In general, an arachnid has a body divided into two sections. The front section consists of a head and a breast (thorax), fused together into one unit – the cephalothorax. The hind section is the abdomen. On true spiders, these two sections are clearly divided by a constriction. Amongst other members of the class, *e.g.* harvestmen and mites, the constriction may be lacking, and the abdomen more or less fused with the cephalothorax.

A typical arachnid has eight walking legs and is easily distinguished from an insect which, of course, can have only six. The four
pairs of walking legs are attached to the cephalothorax.

In addition to the walking legs, there are two pairs of appendages
growing from the front of the cephalothorax. The first pair of these
serve as jaws. These two appendages are comparatively short and
are situated in front of the mouth. They are sometimes compared to
the antennae (feelers) of an insect, but their function is quite different from that of antennae. On the true spiders, for example, they
constitute the poison fangs. The second pair are usually somewhat
similar in appearance to the walking legs, but with fewer joints.
They are quite powerful on a scorpion, the last joint on each being
developed as a pair of pincers.

*Oxygen.* Arachnids obtain oxygen by various methods. Most of
them absorb it entirely from the air by means of book lungs (also
called lung books), tracheal tubes, or a combination of both. There
are some species where these organs are no longer present and
oxygen is absorbed directly through the body wall.

*Food.* Most arachnids feed on other animals – chiefly other arachnids, insects, crustaceans, centipedes and millipedes. Many mites
feed on animal foods, but some feed on plants. Amongst the mites
are found both scavengers and parasites.

*Growth.* Most young arachnids resemble their parents in appearance, although there may be certain differences. Some mites, for
example, have only four or six legs at first. A young arachnid moults
its skin several times during growth.

*Reproduction.* Most arachnids have young by laying eggs. Scorpions
and some mites give birth to young animals which have already
hatched out of their eggs.

In the British Isles the arachnids are represented by spiders,
harvestmen, false scorpions, and mites.

## Spiders

A spider is most easily distinguished from an adult insect by its
number of legs – eight against an insect's six. Other noteworthy
differences are:

1  Most adult insects have wings. No spider has wings.
2  Insects have three parts to their bodies – head, thorax (chest)
   and abdomen. A spider has only two parts – the cephalothorax
   and the abdomen. On the body of a true spider these two
   sections are clearly separated by a waist.

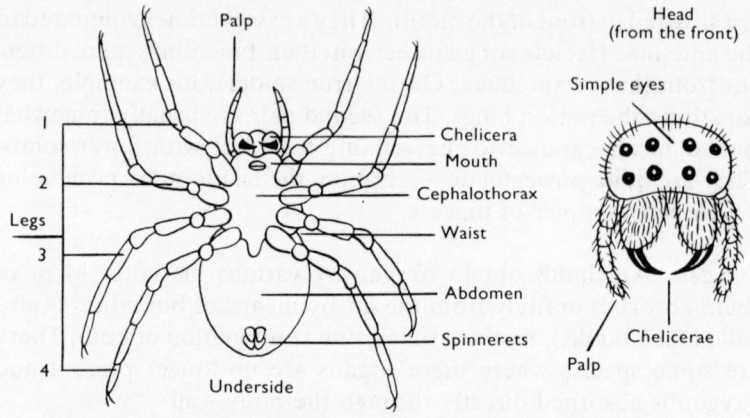

PARTS OF A TYPICAL SPIDER

On the cephalothorax of a spider there are:
*a* four pairs of walking legs;
*b* one pair of palps. These two limbs are organs of touch, and
   certain of their functions are similar to those of the antennae
   of insects;
*c* one pair of chelicerae, serving the purpose of poison fangs;
*d* eyes. These usually number eight, although certain spiders
   have fewer. The eyes are simple, and not compound like those
   of an insect.

On the abdomen there are the short tubular spinnerets from which
silk is extruded. These are usually six in number. Spiders make
continuous use of silk throughout their lives for a variety of purposes. The best-known of these is, of course, for the construction of
some kind of web to trap small animals for food, although there are
many spiders which do not use their silk for this purpose.

   Certain other animals produce silk, *e.g.* caterpillars of moths and
butterflies, but none uses it to the same extent as does a spider.

   From the legs and bodies of spiders grow bristles (setae).
Although hair-like in appearance, these are not true hairs like those
found on a mammal.

*Oxygen.* Oxygen is obtained from the air by means of book lungs, or tracheal tubes, or by the combined use of both. Book lungs enable a greater absorbing surface to be presented to the air. They have been likened to purses with a number of compartments.

*Food.* Spiders feed on other small living animals – mainly insects. They either go out searching for their prey, or they lie in wait for suitable prey to come to them. Of those which go out searching for food, some have no particular home and wander from place to place preying on any suitable animals available. Others take shelter in one place, returning after each expedition.

*Spiders which search for their food (The hunters)*

*a* The zebra and other jumping spiders stalk their prey and make a final spring.
*b* Others, like the wolf spiders, merely pursue the quarry, depending entirely on speed for overtaking it.

*Spiders which lie in wait (The trappers)*

*a* Some, like the crab spiders, hide from view and leap on passers-by. Others have means of camouflage and remain exposed, ready to seize the unwary.
*b* Some – the trapdoor spiders – dig holes in the ground, line them with silk, and close them with silk trapdoors covered with earth. The only British representative of this kind is found in the southern counties. Amongst the trapdoor spiders are the tarantulas, which are the largest spiders known. In the tropics some are able to attack and consume the tiny humming birds, which earns them the name 'bird-eating spiders.'
*c* Some spiders are more advanced in their methods of trapping food and construct special nets – webs – to ensnare their victims.

Not all insects are suitable prey for spiders. The skin of some – for example, weevils and certain other beetles – is sufficiently tough to protect them from spiders. Others, such as the cinnabar moth, aphid, ant and certain flies, are unpalatable to spiders, and if ensnared in a web, are either ignored or cut out from the web. Nevertheless it has been calculated that the total weight of insects consumed by the huge population of spiders in England and Wales is greater than the total weight of the human population of the two countries. This alone is indicative of the considerable value of spiders to human beings.

The poison excreted by the fangs of a spider when it bites its victim is used to kill in some cases, but in others merely to paralyse. Web-spinning spiders generally paralyse their prey in order to render it inactive, but at the same time to keep it alive and fresh. Only a few spiders produce poison which may be dangerous to human beings, the best-known of these being the American black widow spider. The poison of British spiders is harmless to human beings and can cause no more than a temporary irritation.

*Growth.* Spiders moult their skins as they grow. A baby spider hatching from the egg is unable to feed or to spin until after the first moult. After this it resembles its parent, except for colour and size. Baby spiders do not avoid the light as many adults do. They also have a tendency to climb upwards.

The young spiders of many species disperse by aerial migration. A young spider departing in this way climbs to the highest point possible, turns its head to the breeze and exudes a drop of silk, which is drawn out into a thread by the breeze. When this thread, streaming in the breeze, is long enough to exert a pull, the spider releases its grip and is carried away. In this way spiders disperse over hundreds of miles and are carried across the seas to other lands. The threads on which the young spiders are carried are referred to as gossamer and the young spiders are sometimes called 'gossamer spiders'. Gossamer spiders are, of course, not the young of one particular species of spiders, but the young of a number of species.

*Reproduction.* Spiders have young by means of eggs. The female spider lays a cluster of eggs and surrounds them with a protective cocoon of silk. The cocoons of most species are attached to things such as walls, trees or fences and left until the following spring. However, some spiders take care of their young. Some guard the cocoon until the young emerge and may even assist by tearing open the cocoon at a suitable time. The young of some of the web-spinning spiders live for a time in the web together with the parent.

It is not true that female spiders always eat their male partners after mating, but it can happen – especially towards the end of autumn when the females are hungry and the males less active.

There are over 20 000 species of spiders in the world, about 560 of which are found in the British Isles. The largest spiders in the world have a body length of about 7.6 centimetres, and the smallest less than 0.6 millimetres. The largest of the British spiders (the Cardinal spider) has a body length of nearly 2 centimetres, and a leg span of about 12.5 centimetres.

The belief in the curative powers of spiders was once very strong. A very common remedy for fevers in the seventeenth and eighteenth centuries consisted of a house spider rolled in new bread and swallowed as a pill. Less fussy medicals recommended it to be taken spread on a piece of bread and butter.

*Zebra Spider*. This is a hunter, belonging to a family of spiders which capture their prey by approaching slowly and springing suddenly. It is common on walls and fences in the British Isles, and trails a drag-line of silk, which it fastens at intervals to save itself from falling if it misses its footing.

*Wolf Spider*. Wolf spiders are hunters which rely on speed to overtake their prey. They are common in woods, gardens and fields during spring, summer and autumn. The female takes care of her young, carrying the eggs in their silk cocoon attached to her spinnerets. When hatched, the young are carried on the back of the female for a week or two, until they are old enough to disperse by means of gossamer threads.

*House Spider*. Found in sheds, garages and houses, these spiders construct traps. The trap consists of an untidy mat of threads – the familiar cobweb – and the spider lives in a silk tube usually located in one corner of the web. Food includes the disease-carrying housefly.

*Garden Spider*. This belongs to a family which constructs the most advanced type of web – the orb web. This is methodically constructed, consisting of a frame to which are attached a number of spokes radiating from a central hub, and completed by a length of adhesive thread fastened to the spokes in the form of a spiral. The spider may be found either on the web or in a nearby retreat, where vibrations from the web inform it when something has been caught.

Relative sizes of female
and male garden spiders

301

*Freshwater Spider*. The British freshwater spider spins a sheet of silk which is attached to freshwater plants. Air collected at the surface is released below this sheet, which is thus buoyed up into the shape of a bell. The spider carries its own bubble of air about with it. This can be seen as a silvery sheath surrounding the abdomen. When additional air is required for the bell, the spider increases the amount it collects at the surface, by extending its back legs, so that a larger bubble is obtained. Some of this larger bubble is released into the bell. As the oxygen in this store is used, more may diffuse in through the sides of the bell, from the surrounding water.

The bell itself is used solely for dwelling purposes, and not as a trap, as the spider hunts for the various small aquatic animals on which it feeds. These are consumed in the bell or occasionally on land.

## Harvestmen

Harvestmen are also known as harvesters or harvest spiders, as they are very conspicuous during the late summer and early autumn. A harvestman is easily distinguished from a true spider by its long thin legs and its small oval body. The long legs – in some species about twenty times the length of the body – help this arachnid to move quickly and also to raise the body well clear of the ground. The head-chest and abdomen are not clearly separated by a waist as on a true spider, so that the body appears to be a single oval unit. Harvestmen have the same number of limbs as other arachnids, but there are only two eyes. The long legs are shed very readily when grasped, and in consequence it is not uncommon for harvestmen which have been caught by children to be short of their full complement of eight. Unlike the true spiders, they are unable to regenerate these lost legs.

*Oxygen*. Oxygen is obtained from the air by tracheal tubes. Harvestmen do not have book lungs like true spiders.

*Food*. Harvestmen usually avoid sunlight and feed mainly at night. Like the true spiders, they are animal-eaters, feeding chiefly on small spiders, mites and small insects. They are unable to poison their prey but use their fangs to tear it. In captivity they have been observed sampling such fare as fat, meat, bread, butter and syrup. As they are thirsty animals, they are often found near water or in moist surroundings.

*Growth*. Young harvestmen resemble the adults in all except size.

They grow during the spring and summer, casting their skins periodically.

*Reproduction.* Harvestmen mate in the autumn. After the eggs are laid in cracks and holes in the soil, the adults die. The eggs hatch with the warmth of the following spring.

### Other Arachnids

True scorpions and false spiders are not found in the British Isles; they are confined to warmer climates. The true scorpion is the largest of the land-living arachnids, and is especially famous for the sting in its tail. This 'tail' is actually the abdomen, and the sting is used for paralysing prey. Its poison is strong enough to have an effect on human beings. The palps of a scorpion are also well-developed, the last joint on each forming a claw.

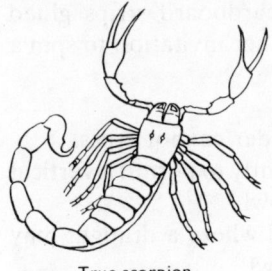

False scorpions and mites are found in the British Isles. These have the usual eight legs, but they are often too small to be observed properly without magnification. Ticks are large mites. The tortoise tick is an example, found with its head buried in the rear limbs of newly-imported tortoises.

True scorpion

### CODE

1 Observe on a true spider
   *a* its eight legs,
   *b* its fangs and feeling limbs,
   *c* its waist.
2 Observe on a harvestman
   *a* its eight legs (remembering that harvestmen are not infrequently found with one or more legs missing)
   *b* the fact that it has no waist.
3 Observe the various uses to which spiders put silk threads.
4 Observe any collected land spiders and harvestmen in containers similar to those used for insects, but shade them from direct light. House and garden spiders in particular may be fed on houseflies or bluebottles. Adult bluebottles may be obtained by leaving two or three maggots to pupate in the cage. The introduction of soil and grass should ensure a limited supply of small

animals, suitable for food. As harvestmen are thirsty animals, a small pot of water should be provided for them. As spiders feed on other animals they are difficult to keep for any length of time, unless suitable food can be provided. However, provided they are well fed to begin with, spiders are, in general, able to withstand fairly long periods without food. Nevertheless it is kinder to release captive spiders when the holidays arrive.

5   Observe any colleced freshwater spiders for a while in small aquaria or toffee jars. Pond water containing small animals is better than tap water. The addition of some freshwater herbs may encourage a freshwater spider to construct a nest.

6   Observe an orb web spider by confining it to a spider island. Not many spiders can travel across the surface of water so that an orb web spider may be encouraged to make its home on a potted plant by isolating it on an inverted tin or plant pot standing in a bowl of water. A rough polygon of cardboard strips glued together at the ends may serve as a further invitation to spin a web.

*Notes*

*a* The water provides a moat which the spider cannot cross.

*b* No part of the plant should extend sufficiently to permit a vertical drop to be made to the outside of the bowl.

*c* The 'spider island' should not be situated where a draught may carry a thread beyond the wall of the bowl.

*d* The inverted tin is to raise the plant-pot clear of water.

*e* The rough polygon of cardboard is an invitation to spin which is not always accepted.

TEMPORARY HOMES FOR LAND SPIDERS

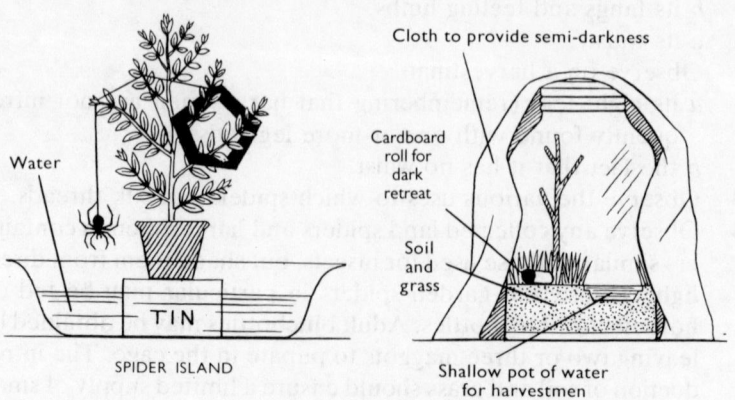

Cloth to provide semi-darkness

Cardboard roll for dark retreat

Soil and grass

Water

TIN

SPIDER ISLAND

Shallow pot of water for harvestmen

7   Observe a house spider and its cobweb where perchance it makes its home. Cobweb spiders have been known to make their homes uninvited in the classroom. Such spiders are useful for observation purposes if the cleaners can be coaxed into leaving them alone.

8   Preserve required specimens in 5% formaldehyde in the usual way.

## Written Work

1   True spiders and harvestmen have eight legs, two fangs, and two feeling limbs.

2   True spiders have waists. They can spin silk.

3   Some kinds hunt their prey. Some hide and wait.

4   A harvestman has no waist and spins no silk.

## 35   ANIMALS WITH CRUSTY SKINS

### CRUSTACEANS

### Demonstration Material

Any of the following:

1   Living or preserved specimen of a marine crustacean, *e.g.* lobsters, crab, prawn, sea shrimp.

2   Living or preserved specimen of a freshwater crustacean, *e.g* crayfish, freshwater shrimp, freshwater louse.

3   Living or preserved specimen of a woodlouse.

4   A dead part from a crustacean, *e.g.* a claw, or part of the crusty skin of a crab or lobster.

### Sample Link Questions

1   How many legs has a spider or a harvestman? (*Eight*)

2　What does a spider or a harvestman have on the front of its body besides a pair of feeling limbs? (*A pair of fangs*)

3　What divides the body of a true spider into two parts? (*A waist*)

4　How does the body of a harvestman differ from the body of a true spider? (*It has no waist*)

5　Which spiders make webs of silk to trap their prey – those which hunt or those which hide and wait? (*Those which hide and wait*)

6　Do most kinds of spiders live in water or on land? (*On land*)

7　What do fish have to help them take oxygen from the water? (*Gills*)

8　Which two classes of animals have wings? (*Birds and insects*)

9　How many legs has an adult insect? (*Six*)

10　How many feelers has an insect? (*Two*)

**Relevant Information**

The main purpose of this lesson is to introduce the class of animals with crusty skins – the *Crustaceans*.

1　They have two pairs of feelers and more than six legs.

2　Most kinds live in sea water, some kinds live in fresh water, and very few live on land.

The four groups of animals which have jointed limbs but no internal skeleton (*i.e.* the insects, arachnids, crustaceans, and the centipedes and millepedes) have a tough outer skin (the exoskeleton) which serves to protect the soft inner parts of these animals against enemies. This outer skin is made of a horny substance called chitin. It is particularly apparent on crustaceans, where a very efficient crust results. This hard crust has been likened to a suit of armour and serves a similar purpose.

Crustaceans vary greatly from one another in size, shape, and number of limbs. In general they are aquatic animals, living in sea water or fresh water, absorbing their oxygen from that dissolved in water, and feeding on living or dead things found there. In this respect they differ from the vast majority of insects, the arachnids, and the centipedes and millepedes, which are typically land animals. But just as a small number of insects and arachnids are aquatic, so a small number of crustaceans have adapted themselves to living and feeding on the land. Such crustaceans include the woodlice and the land crabs, which obtain their oxygen from the air. Woodlice are very common in the British Isles and are completely adapted to a terrestrial existence. They even breed on land – unlike the land crabs of tropical countries which have to return to the sea to breed, and which spend the first part of their lives in water.

But although a few crustaceans are found on land and quite a number in fresh water, salt water is the domain of the vast majority. These include crabs, lobsters, prawns, shrimps, barnacles and a great variety of very small crustaceans. Many are microscopic in size and form part of the plankton floating on the surface of the sea.

*Oxygen* is absorbed from the water either by means of gills or, in the case of very small crustaceans, directly through the surface of the body. Very few obtain it from the air.

*Food* varies considerably; some feed on living animals or plants, others on dead animals or plants. Some of the small crustaceans are parasites.

*Growth.* The majority of crustaceans hatch from eggs looking quite different from the adults, and pass through a rather complicated metamorphosis. The free-swimming young are known as larvae. Certain crustaceans, *e.g.* woodlice, freshwater shrimps and sand-hoppers, emerge from their eggs resembling their parents in appearance.

Owing to the nature of the hard protective crust which covers it, a crustacean is forced to moult its skin periodically in order to grow. When the old skin has been shed, the owner is quite defenceless until the soft new skin has hardened. It is during this period between the shedding of the old skin and the hardening of the new one that the actual growth takes place. A considerable number of moults may be necessary. It has been estimated, for example, that a young lobster moults about 14 to 17 times during the first year of its life, although it may reach a length of only 5 to 7 centimetres during that time.

*Reproduction* is usually by means of eggs. Only very few species retain eggs in the body until they have been hatched. On the other hand, many crustaceans do carry the eggs about with them, attached to some part of the body, until they have hatched. Woodlice, freshwater shrimps and sandhoppers have a special brood pouch for this purpose.

A characteristic of a crustacean is the two pairs of feelers growing from the head. These are particularly noticeable on lobsters, prawns, sea shrimps and crayfish. Limbs are in pairs, and the number varies from species to species. The lobster, prawn and sea shrimp are typical of the more complex crustaceans, on which the limbs have become specialised. The limbs nearest the mouth have become specialised for feeding, those farther along the body for walking, and those nearest the tail – the swimmerets – for swimming.

The six crustaceans illustrated in the Pupils' Book are all malacostraca – one of the five major sub-classes into which the crustaceans are divided. This sub-class contains most of the crustaceans large enough to be commonly noticed by children.

*The Common Lobster* has its first three pairs of legs equipped with claws. The front pair consists of a large claw with blunt knobs for crushing, coupled with a smaller one with saw-edged blades for holding and tearing. Food consists of dead animals or of living animals such as fish and clams. Oxygen is obtained by forcing water over feathery gills attached to the bases of the legs. Eggs are carried adhering to the female's swimmerets. When hatched, the young lobsters are free-swimming and pass through a series of stages before reaching the adult form. The adult inhabits shallow waters off rocky coasts, crawling about in search of food, and propelling itself backwards with its broad tail fan when alarmed. Each of the two eyes is situated at the end of a stalk. Limbs, when damaged, may be discarded, and new ones grown – a characteristic of many of the larger crustaceans. When a lobster is alive, its colour is blue-black with red or orange underparts, but when boiled, it becomes red all over.

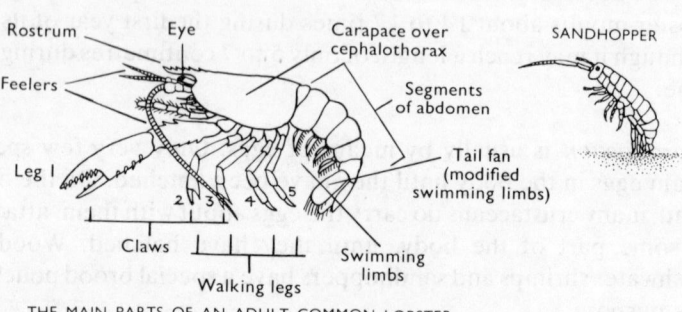

THE MAIN PARTS OF AN ADULT COMMON LOBSTER
(Appendages on one side only are shown)

*The Common Prawn*, though smaller, is similar in appearance to the lobster, but it is a more active swimmer. It has five pairs of walking legs, with pincer-claws on the first and second pairs, and swimming limbs on the abdomen. There is a pair of long feelers, and a pair of short feelers. Each short feeler bears three branches. Oxygen is obtained by means of gills, and food by means of scavenging. Eggs are carried adhering to the swimming limbs and protected by a brood pouch. The young, when they emerge, are free-swimming and pass through several stages before becoming adults. The adults

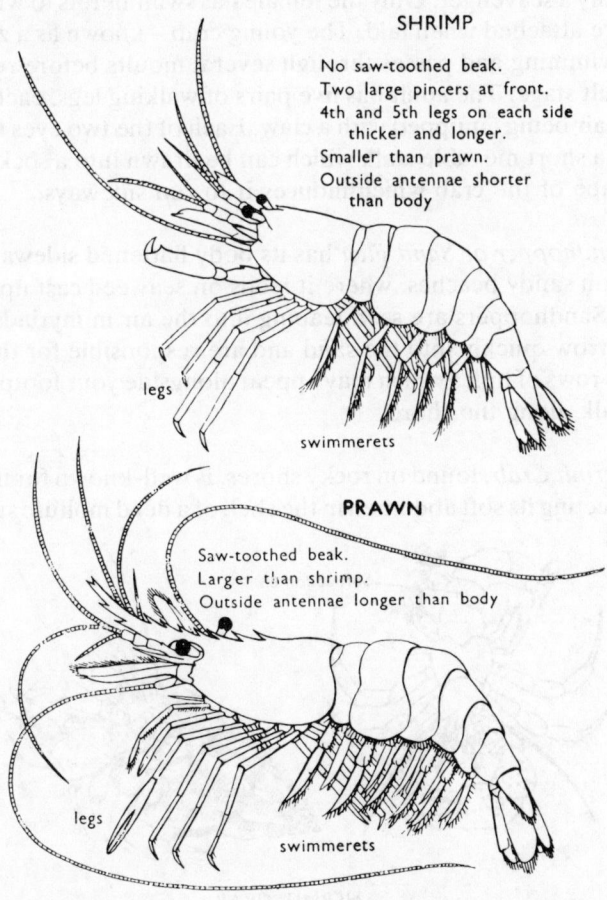

SHRIMP

No saw-toothed beak.
Two large pincers at front.
4th and 5th legs on each side
    thicker and longer.
Smaller than prawn.
Outside antennae shorter
    than body

legs

swimmerets

PRAWN

Saw-toothed beak.
Larger than shrimp.
Outside antennae longer than body

legs

swimmerets

Shrimps and prawns, like crabs and lobsters, are crustaceans with five pairs of legs (decapods). Some of these legs are modified into nippers. On the prawn, the pair of hairy appendages that look as if they were the foremost pair of legs are in fact mouth-parts.

live in shoals off rocky coasts, and are difficult to see, even in sandy pools. It is their eyes which betray them. Prawns are distinguished from shrimps by the saw-toothed spike projecting forwards from the edge of the carapace just between the eyes.

*The Edible Crab*, like most crabs, is distinguished from lobsters, prawns and shrimps by having the abdomen reduced to a tail-flap tucked up under the thorax. Oxygen is obtained by means of gills. Food consists mainly of living or dead animals, the crab being generally a scavenger. Only the female has swimmerets to which the eggs are attached when laid. The young crab – known as a zœa – is free-swimming and passes through several moults before reaching the adult stage. The adult has five pairs of walking legs, each of the front pair being equipped with a claw. Each of the two eyes is at the end of a short movable stalk which can be drawn into a socket. It is the shape of the crab which induces it to run sideways.

*The Sandhopper or Sand Flea* has its body flattened sideways. It is found on sandy beaches, where it feeds on seaweed cast up on the shore. Sandhoppers are seen leaping into the air in myriads. They can burrow quickly into the sand and are responsible for the mysterious rows of holes which may appear alongside your footprints as you walk along the shore.

*The Hermit Crab*, found on rocky shores, is well-known for its habit of protecting its soft abdomen in the shell of a dead mollusc such as a

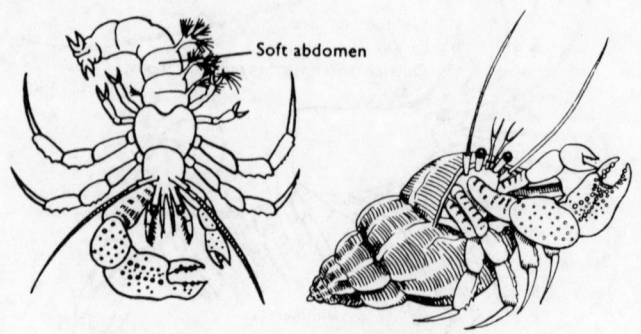

— Soft abdomen

HERMIT CRAB
A crustacean which uses the shell of a dead mollusc (*e.g.* whelk) for protection. As it grows, it transfers to a larger shell.

periwinkle or whelk. Its front two limbs are equipped with large claws of unequal size. As it grows larger, it has to transfer itself to another suitable empty shell.

*Acorn Barnacles*. In spite of their appearance, barnacles are crustaceans but of a lower sub-group. They are free-swimming when first hatched from the egg, but eventually settle down for life on objects such as rocks, shells and piers. A barnacle fastens itself down by its head and kicks food into its mouth with its fringed limbs. It was at one time considered to be a mollusc because of the considerable hardness of its skin. In the Pupils' Book crenated barnacles are shown on the edible crab.

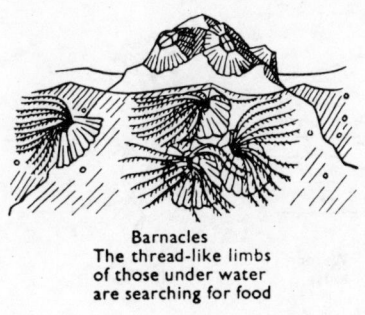

Barnacles
The thread-like limbs
of those under water
are searching for food

*The Common Crayfish* is of course not a fish. It is similar to the lobster in appearance (but smaller), and lives in English and Irish streams, but not usually in those of Scotland. Oxygen is obtained through feathery gills, and food consists of other living animals and occasionally dead things. The young which emerge from the eggs remain attached to the swimmerets of the mother until independent. The adult is about 10 centimetres long and hides in burrows and crevices in the banks of its stream.

*The Freshwater Shrimp* has a body flattened from side to side like a sandhopper. It swims on its side close to the bottom, and is found in shallow running water or still water where there is plenty of oxygen. Oxygen is obtained by means of gills, the first three pairs of limbs vibrating continually to force a stream of water over them. Food is obtained chiefly by scavenging on dead animal or plant parts. Eggs are carried by the female in a brood pouch until they hatch; the young shrimps are similar to the parent in appearance.

311

When these animals are breeding during spring and summer, the male may often be seen carrying the female about.

*Note*

Also common in fresh water are the freshwater louse and freshwater fleas. The *Freshwater Louse* is very similar to the woodlouse, but the legs and feelers are longer. Oxygen is obtained by means of gills. Food consists chiefly of dead animal or plant parts. *Daphnia* and *Cyclops* are two of the commonest of the relatively small freshwater crustaceans known collectively as water fleas. They are smaller in area than the end of a pencil and may often be seen jerking their way through water which is rich in algae

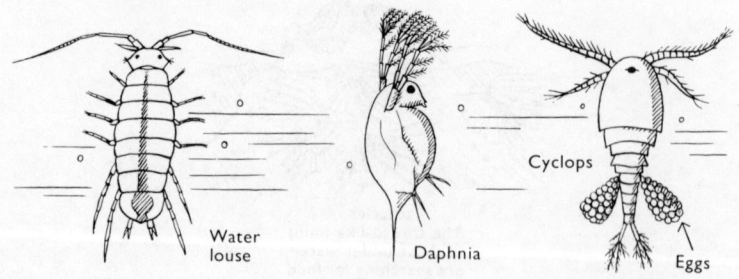

Water louse     Daphnia     Cyclops     Eggs

*The Common Woodlouse*. Woodlice are the only crustaceans which have completely adapted themselves to life on land. They are very common and are found in gardens and on waste land, usually in damp places under stones, bricks and pieces of timber. They are also known as 'slaters'.

They take oxygen from the air by the beginnings of a simple tracheal tube system somewhat similar to that of insects, but located in two pairs of rear legs. Food varies. Woodlice feed mainly on dead plant parts, but they will also feed on living plants and on dead or living animals. They will even attack one of their own kind when it has just moulted its skin.

Woodlice have young by means of eggs, and these, when laid, are deposited and carried in a brood chamber until hatched. The adult is less than 2.5 centimetres in length. A woodlouse has seven pairs of walking legs. The commonest species in British gardens are the brown woodlouse and the grey woodlouse. Another species, the common pill woodlouse, is able to expose itself to the sun more than the other two. It curls into a ball when alarmed, and its name comes from the fact that it was once considered to have medicinal value and was swallowed as a ready-made pill.

## CODE

1 Observe:
   a the hard crusty skin on a lobster, crab, or part of one of these
   b the long feelers on a lobster, prawn or sea shrimp
   c the strong claws for feeding on a lobster or crab
   d that on a lobster, prawn or sea shrimp there are walking limbs and swimming limbs, and that on a crab there are walking limbs but no swimming limbs
   e the prominent 'eyes on stalks' on a lobster, prawn or sea shrimp.
2 Observe:
   a that the freshwater shrimp can move from place to place by swimming
   b that the freshwater louse moves from place to place by walking.
3 Observe that the woodlouse lives on land and usually hides from strong light.
4 Collect marine crustaceans from sea-washed pools on the shore. They are difficult to keep in captivity, unless they can be provided with well-aerated sea water. Children should be encouraged to return them to the pools in which they found them and not to keep them in buckets, where they will die through lack of oxygen.
5 Collect freshwater crustaceans from slow streams, ponds and canals. If rooted freshwater herbs are pulled out of the water and spread out on a sheet of newspaper, freshwater shrimps and freshwater lice may be seen struggling in the mud clinging to the roots. They may be kept in an aquarium containing some of the herbs and mud in which they were found. Freshwater starwort and Canadian pondweed are two plants on which they have been known to feed.
   Freshwater crayfish seem rarer than they were and they are difficult to keep in captivity. One may survive for a while in well-aerated shallow water on a diet of earthworms.
6 Collect woodlice from dark damp situations, e.g. from under rocks, bricks, lengths of discarded timber. They may be kept in a

small tank or large jar containing a mixture of moist peat, pieces of rotting wood, and dead leaves. It is advisable to have a loose cover on the container to maintain a damp environment. If a piece of slate or a flat piece of rock is included, the woodlice will often congregate beneath it.

*a* Preserve aquatic specimens in 10% formaldehyde

*b* Preserve woodlice in 10% formaldehyde.

*c* Parts of a lobster or crab, *e.g.* legs and claws, may be kept without liquid preservative.

## Written Work

1 A <u>crustacean</u> has two pairs of feelers. It has to shed its crusty <u>skin</u> before it can <u>grow bigger</u>.
2 Most crustaceans live in <u>water</u> and breathe through <u>gills</u>.
3 The <u>woodlouse</u> lives on land. It takes oxygen from the <u>air</u> and feeds mainly on <u>dead</u> things.

# QUESTIONS ON LESSONS 25 TO 35

1 What are the three forms in which we find never-alive things? *Solid, liquid, gas*

2 What are the four main needs of living things? *Oxygen, food, to grow, to have young*

3 Of the simple plants, herbs, trees and shrubs, which are the ones that can never have flowers? *Simple plants*

4 From what part on a stem does a flower grow? *A bud*

5 On some plants each flower grows on its own; on other plants flowers grow together in groups.
What do we call these groups? *Clusters*

6 What do we call the separate parts into which a compound leaf is divided? *Leaflets*

7 On one kind of compound leaf the leaflets

314

grow from the side of the leaf stalk. From which part of the stalk do the leaflets grow on the other kind of compound leaf? *The end*

8 Which part usually grows towards the light – a stem or a root? *A stem*

9 What is the special name for the pupa of a moth or butterfly? *Chrysalis*

10 How many true legs does a caterpillar have? *Six*

11 Which has a knob on the end of each feeler – a moth or a butterfly? *A butterfly*

12 Which rests with its wings folded up – a moth or a butterfly? *A butterfly*

13 In which stage of its life is a two-winged fly sometimes called a maggot? *Larva(l)*

14 What do many of the two-winged flies carry which makes them a nuisance to other animals? *Diseases*

15 How many legs does a two-winged fly have when it is a larva? *None*

16 Where do gnats and midges spend the first three stages of their lives? *In water*

17 Does a planet go round a star or round a satellite? *Round a star*

18 What goes round a planet? *A satellite*

19 Which star is nearest to the earth? *Sun*

20 What is the name of the planet nearest to the sun? *Mercury*

21 Is Mars a star, a planet or a satellite? *A planet*

22 What is the main purpose for which dead things are used by living things? *Food*

23 Some animals use dead things for protection. What two things may they need to protect themselves against? *Enemies and weather*

24 Which are the only animals to use dead things for other purposes, as well as for food and for protection? *Human beings*

25 What have the sands of the desert and the shore been worn down from? *Rocks*

26 What are most rocks a mixture of? *Solids*

27 What are the two things which, when moving, wear down rocks? *Air and water*

28 Is it air or water that carries pieces of rock towards the sea?   *Water*

29 Is it air or water that carries the sand of the desert from place to place?   *Air*

30 Is it air or water that wears down the pebbles on the beach?   *Water*

31 What forces the waves of the sea against the shore?   *Moving air (wind)*

32 What do we call herbs which have their roots in mud and their stems and leaves growing out of the water?   *Swamp herbs*

33 What do floating leaves take oxygen from?   *The air*

34 On which class of animals do spiders and harvestmen mainly feed?   *Insects*

35 What is the name for the case of silk which a female spider spins round her eggs?   *Cocoon*

36 Where does the harvestman lay its eggs?   *In the ground (in a hole)*

37 What name do we give to the net of silk which some spiders spin in order to trap their prey?   *Web*

38 How many feelers has a crustacean?   *Four*

39 A crustacean has to shed its crusty skin so that it can do what?   *Grow bigger*

40 What do most crustaceans have to help them to take oxygen from the water?   *Gills*

 **36** **LAND ANIMALS WITH MANY LEGS**

## CENTIPEDES AND MILLEPEDES

### Demonstration Material

1 Any living or preserved centipede.
2 Any living or preserved millepede.

### Sample Link Questions

1 What name do we give to animals with crusty skins? (*Crustaceans*)

2   Which can have wings – insects, spiders or crustaceans? (*Insects*)
3   Is a spider's body divided into two parts by a waist? (*Yes*)
4   Where do most kinds of crustaceans live – in water or on land? (*In water*)
5   Name a common crustacean which lives on land and takes its oxygen from the air. (*Woodlouse*)
6   How many feelers has a crustacean? (*Four* or *two pairs*)
7   How many feelers has an insect? (*Two*)
8   How many legs has an adult insect? (*Six*)
9   How many legs has a true spider or a harvestman? (*Eight*)
10  Some of the limbs on a lobster and prawn are used for walking and feeding. What are their other limbs used for? (*Swimming*)

**Relevant Information**

The main purpose of this lesson is to introduce the class of animals with many legs, *i.e.* the centipedes and millepedes.

1   Centipedes and millepedes are land animals. None live in fresh water or sea water.
2   A centipede has one pair of legs on most of its body segments. A millepede has two pairs of legs on most of its body segments.

The centipedes and millepedes belong to a group of animals known collectively as the *Myriapoda*. Together with the insect class, the arachnid class and the crustacean class, they are animals with jointed limbs and external skeletons.

Generally speaking these animals are easily distinguished from insects, arachnids and crustaceans, by their long narrow bodies and their greater number of segments. They are also conspicuous by their many walking legs, which are located all the way along the body.

The head of a centipede or millepede has one pair of feelers (antennae), and the body is clearly divided into segments, the number of which varies from species to species. The body colour is generally fawn, brown or black, although there are some variations. The body length also varies from species to species. Some are so tiny as to be hardly visible, and others may exceed 25 centimetres in length.

Oxygen is obtained from the air by means of air holes and a system of tracheal tubes similar to that of insects. Most centipedes and millepedes feed at night and hide during the day in dark places, *e.g.* under stones and lengths of discarded timber.

Reproduction is by means of eggs, and in a number of species the young lack the full complement of segments when they first hatch.

There are thousands of species of centipedes and millipedes distributed throughout the world. They are all land animals, although a few species live on the shore between tide marks and are able to survive fairly long periods of immersion under water.

## Centipedes

The popular idea that a centipede has 100 legs is of course false. Some have less, and some have more. Centipedes are easily distinguished from millepedes by having only one pair of walking legs to each segment of the body. In general, the feelers on the head of a centipede are longer than those on the head of a millepede. Also, the body of a centipede is more flattened, while that of a millepede is more rounded. Some centipedes have short squat bodies, but others have bodies which are comparatively very long and thin.

Centipedes feed on other living animals, and each of the first pair of legs is equipped with a poison claw to assist in killing prey. However, although centipedes are generally accepted as being animal-eaters, they may not all be completely carnivorous, as cases have been reported of centipedes feeding on plants such as celery, lettuce and onion.

Most centipedes have eyes, but on some species they are absent. The young of some species of centipedes leave the eggs equipped with the full number of walking legs and segments. The young of other species hatch from the eggs equipped with only seven pairs of legs, including the first pair which has the poison claws. These centipedes develop further segments and walking legs as they grow larger.

British centipedes are useful animals to the gardener because they feed on slugs and various insect larvae. It should not be assumed that they discriminate, however, as they feed on insects which are useful to the gardener as well as on those which are a nuisance.

*The Common Centipede.* This is the common brown centipede of our gardens. The head is noticeable, and the feelers are long and

318

conspicuous. On each side of the head, there is a group of four eyes. The segments of the body are alternately small and large, and there are fifteen pairs of legs. The common centipede hides during the day, often under leaves or heaps of garden rubbish, and emerges to hunt for its food at night. During the winter it buries itself in the earth and rests.

The eggs are laid from June to August, and each egg, as it is laid, is covered by a sticky liquid. As an egg appears, the female rolls it in soil, which adheres to it and serves to camouflage it. This precaution is necessary to save it from the attentions of the male centipede who will devour it eagerly if the opportunity arises.

The young common centipede which hatches out of the egg has only seven pairs of legs – including the first pair bearing the poison claws. As it grows, additional segments with legs make their appearance just in front of the rear segment.

Other names for the common centipede are 'stone-dweller' and 'thirty legs'.

*The Earth Lover*. The earth lovers are the most powerful British centipedes. They have conspicuously long bodies which are usually pale reddish-brown in colour. The segments are all about the same size and number from 31 to 173. The legs are shorter than those of the common centipede.

During the day the earth lover hides, usually amongst moss or rotting vegetation, and emerges at night to hunt for its food. Unlike the common centipede, it has no eyes.

Earth lovers hatch from their eggs with a full complement of segments and legs. These centipedes are widely distributed throughout Britain but are more common in the south of England.

A species of earth lover found in Cambridgeshire and in Epping Forest (*Geophilus electricus*) is phosphorescent at night, emitting a glow like that of a glow-worm.

## Millepedes

Millepedes are easily distinguished from centipedes, as they have two pairs of legs on most of the segments of the body. The first few segments behind the head are generally exceptional. Among these are usually found three segments with but a single pair of legs each and at least one segment with no legs. The total number of legs varies. Large millepedes may have as many as 200 pairs.

In general, the two feelers on the head of a millepede are shorter than those on a centipede. As they are short and clubbed and often bent at an angle, they are not always noticeable.

The bodies of millepedes present a variety of shapes, but the general tendency is towards a tubular shape. Body length also varies from species to species. Some are less than 4 millimetres long, and some – found in the tropics – can exceed 20 centimetres in length. The movement of a millepede from place to place is not so rapid as that of a centipede. On the whole it is more graceful, and waves of movement seem to pass along the animal as it progresses.

Millepedes feed exclusively on plants. Some feed on the softer tissues of living plants, and others on dead or decaying plant parts. As they are plant-eaters, they are not equipped with poison claws. In this respect they again differ from the centipedes.

There are some millepedes, living habitually in dark places, that are without eyes, but most millepedes have them. Their eyes are always simple, and never compound like those of an insect. When alarmed, millepedes curl up, and many are also furnished with stink glands which can exude a liquid with a pungent smell.

Some millepedes make special nests in which to lay their eggs. Others merely deposit groups of eggs either in soil or in decaying plants. Information on their life history is not by any means complete, however.

It is because some millepedes feed on living plants that gardeners generally consider them to be pests.

*The Common Millepede.* This is the most common of the British millepedes. It is a small dark animal, found in gardens and among cultivated crops, and is sometimes known as the 'false wireworm'. (The term *wireworm* is also given to the larva of the click beetle.) The head of this millepede is fairly conspicuous and bears the eyes in front and the feelers at the sides. The body is long and tubular and in the adult stage has more than thirty segments. When alarmed, the common millepede rolls into a spiral and may emit an evil-smelling liquid from the stink glands which are arranged along both sides of the body. It sometimes bores into the underground part of a cultivated plant.

Eggs, numbering from 60 to 100, are laid from May to July. The female prepares a small spherical nest a little way below the surface of the ground before laying the eggs. The young millepedes hatch in about twelve days under normal conditions. In its first walking phase, the young millepede has only three pairs of legs. As it grows bigger, it develops new segments between the penultimate and the last segment. These segments usually come in sets of five, and later develop legs, so that the legs of this millepede appear in batches as it grows.

*The Pill Millepede*. This common British millepede bears more resemblance to a woodlouse than to other millepedes. The body is shinier than that of a woodlouse, however, and of course bears more legs. When alarmed, the pill millepede rolls itself into a ball for protection.

Eggs are laid in the soft or loose soil under moss or dead leaves. After it has been laid, each egg is plastered with an excreted fluid which hardens into a protective case. These eggs in their cases are deposited a few at a time in separate groups.

In its first walking phase, a young pill millepede has only three pairs of legs. As it grows bigger, it develops additional segments and legs with each moult.

The process of moulting usually takes place beneath the doubtful protection of moss or dead leaves. As with most millepedes, the moulted skin is usually devoured by its owner.

## CODE

1    Observe on a centipede or millepede that walking legs are located all the way along the body.

2    Observe on a centipede:
*a* the one pair of legs to each segment
*b* that the body is more flattened than that of a millepede
*c* that the feelers are prominent.

3    Observe on a millepede:
*a* the two pairs of legs to almost every segment – so closely packed that they are difficult to count
*b* that the body is more tubular in shape than that of a centipede
*c* that the feelers are difficult to see.

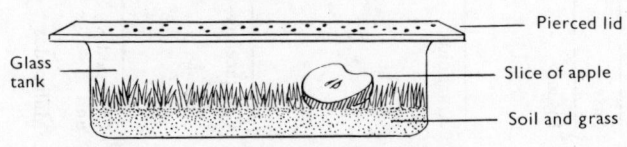

A HOME FOR CENTIPEDES AND MILLEPEDES

A summary of the main observable differences between the adults of the principal classes of animals with jointed limbs and no backbones

| Class | Walking Legs | Head and Body | Wings | Antennae | Where found |
|---|---|---|---|---|---|
| Insects | Always 6 | Separate head, thorax and abdomen | Most have 2 pairs; some have 1 pair; a few have none | 1 pair | Mostly on land; some in fresh water |
| Arachnids | Always 8 | Head and thorax fused into one unit (cephalothorax). Abdomen may be clearly distinct as on waisted spiders; sometimes indistinct as on harvestman, mites etc | None | No true feelers. 2 pairs of appendages grow from front of cephalothorax | Mostly on land; some in fresh water; some in the sea |
| Crustaceans | Number varies. E.g. 5 pairs on lobster; 7 pairs on woodlouse | Head, thorax and abdomen not generally distinct | None | 2 pairs | Mostly in sea water; some in fresh water; woodlouse on land |
| Centipedes and | All the way along the body. One pair per segment | Head distinct. Body divided into segments | None | 1 pair (long) | On land |
| Millepedes | All the way along the body. Two pairs on nearly every segment | Head distinct. Body divided into segments | None | 1 pair (short) | On land |

*Note*

In captivity centipedes and millepedes are not very rewarding animals, as they disappear from view when given the opportunity. If it is desired to keep living specimens, they may be confined in a small glass tank or jar containing a layer of soil and grass. The centipedes, being animal-eaters, are the most difficult to cater for, once they have consumed any slugs and insects found in the soil. The millepedes may be induced to feed on a slice of apple and on leaves.

4 Preserve required specimens in 10% formaldehyde in the usual way.

**Written Work**

1 Centipedes and millepedes take oxygen from the air. They have young by means of eggs.
2 A centipede has one pair of legs on most of its segments. Centipedes feed mainly on other animals.
3 A millepede has two pairs of legs on most of its segments. It feeds on plants.

# INDEX

Achene 66, 67
Air 40
  and weather 109–10
  as carrying agent 92–4
Algae 53, 55, 212, 213
  in fresh water 293
'Alive' 34–5
Alligators 203
Amphibians 88, 184–97
  evolution of 185–6, 188
  in captivity 194–7
  respiration 157, 187
  size 187
Animals 69
  as carrying agents 93
  care of young 180–1
  food of 86–91, 276
  movement of 86–91
  reproduction 176–80
  self-protection 276–7
  simple 71–2
  size 48
Ants 74, 77, 78, 79–80, 181, 276
Apollo missions 267
Aqueducts 142
Arachnids 296–303, 306, 322
Asteroids 270

Bacteria 54, 86, 274, 275, 293
Barnacles 311
Berry 64
Birds 88, 89, 90, 184
  care of young 180–1, 276
  evolution of 202
  respiration 157
  size 48
Buds 171–3
  adventitious 171
  axillary 171
  lateral 171
  scales 173
  structure 172
  terminal 171–2
Bulbs 237, 238
Burning 158–9
Butterflies 77, 245–9
  eggs 244, 249
  larvae 245–6, 248, 249
  pupae 247, 248–9

Capsules, Egg 178, 183; Seed 65
Carpels 62–3
  in dry fruits 65
Carrying 97
Caryopsis 66
Caterpillars 76, 245–6
  keeping 83–4
  looper 246

Centipedes 316–19, 323
Chlorophyll 47–8, 229
Clouds 110, 138, 146
Clubmoss 54, 58, 214
Comets 270
Cones 58, 172, 214
Corms 237, 238
Crabs 310
Crayfish 311
Crocodiles 203
Crustaceans 306–13, 322

'Dead' 34, 35–6
  material, use of 274–8
Dinosaurs 200, 201
Dissolving see Solutions
Drupe 64–5

Earth 265
  crust of 280
  erosion of 280–4
  pull of 114–21
Earwigs 74, 78
Egg capsules 178, 183
Eggs 48, 177–9
  amphibians' 187, 188
  birds' 178
  crustaceans' 307
  fish 178
  insects' 74
  reptiles' 203, 204, 205
  spiders' 300
Emulsions 127–8
Endocarp 64
  in drupes 64
Epicarp 64

Ferns 54, 58, 214
Fish
  reproduction 178, 180
  respiration 157
  size 48
Flies 76, 77, 252–6
  eggs 252
  larvae 252
  pupae 253
Florets 225, 226
Flowers 59–60, 214, 223–7
  composite 225–6
  in clusters 225
  on their own 225
  parts 61–3
  sessile 225
Fluid 42
Foam 132
Fog 110–11, 129
Follicle 65

324

Food
  dead material as  274–6
  of animals and plants  47–8, 138
  of butterflies  247–8
  of flies  253
  of harvestmen  302
  of insects  77, 82–4
  of larvae  76, 246, 252
  of moths  247–8, 276
  of spiders  297, 299
Freezing  145–9
Friction  97, 98–100
Frogs see Amphibians
Frost  147
Fruits  60, 63–7, 214
  accessory  67
  adapted for carriage  94
  aggregate  67
  composite  67
  dehiscent  65–6
  false  61, 67
  indehiscent  65, 66
  simple  64–6
  true  61
Fungi  54, 56, 86, 212, 213, 215, 217,
    275
  in fresh water  292, 293

Gases  40–41
  in solutions  132–3, 134
  mixtures containing  126–7, 128–9,
    132
Gills  157
  tracheal  158, 297, 299
Glaciers  149
  erosion by  283
Gravity, Force of  97, 114–21, 138
Growth  48
  of crustaceans  307
  of harvestmen  302
  of spiders  300
  of stems and buds  171–3

Hail  111, 147
Harvestmen  302–3
Herbs  52–3, 54
  aquatic  287–92
  floating  290–1
  flowers on  223–7
  leaves on  229–32
  reproduction  222–3
  stems  236–7
    and buds  171–3
  swamp  288
  underwater  291–2
Horsetails  58, 214
Hurricane  109

Ice  146, 148
Inflorescence  223–7
Insects  72–80, 322

Insects contd.
  cages for  83
  food of  77, 82–4
  respiration  157–8
  skin  306
  size  48, 77
  social  181
  with a four-stage life  72–7, 78
  with a three-stage life  73, 74, 78–9

Ladybirds  78, 79
Larvae  74–6
  food of  74, 76, 82–4
Leaves  229–32
  alternate  230
  compound  229, 231
  functions of  229–30
  modified  237
  opposite  230
  sessile  231
  simple  229, 231
  whorled  230–1
Legume  65
Liquids  39–40
  as solvents  132–4
  mixtures containing  127–8, 132
Liverwort  54, 213
  in fresh water  292, 293
Lizards  199, 204–5
Lobsters  308
Lungs  88, 157, 199
  book  297, 299

Magnetism  163–5
Magnets
  care of  166–7
  making  165–6
  natural  164–5
Mammals
  care of young  180–81
  reproduction  179–80
  respiration  157
Mass  119–20
Materials (equipment)  30–32
Melting  134, 145
  ice and snow  148–9
Mesocarp  64
Meteors  270
Millepedes  306, 317–18, 319–21
Mist  111
Mixtures  125–9
  colloidal  126
  emulsions  127–8
  molecular  127, 133
  solutions  126, 127
Monsters, Prehistoric  200–02
Moss  54, 56, 213
  in fresh water  293
Moths  73, 76, 77, 242–9
Mounting specimens
  methods of  23–7
  materials for  33–4

'Never-alive'  34, 36–7
Newts *see* Amphibians
Nut  66

Osmosis  238
Ovaries, Plant  62–3
Ovules  62, 63
Oxygen  46–7, 82, 86, 88, 156–60
  and burning  158–9
  and respiration  157–8

Parasites  54, 238, 239–40, 275, 307
Pastes  132
Perianth  62
Pericarp  64
  in dry fruits  65–6
Petals  62
Photosynthesis  47, 229
Pistils  62–3, 67
Planets  260, 261, 262, 264–5, 268–70
Plants  51–6
  as food for animals  86
  flowers' on  222–7
  food of  86, 229, 274–5
  in fresh water  287–93
  leaves on  229–32
  reproduction  211–6, 222–3
  roots of  237–41
  size  48
  stems of  236–7
Pod  65
Pome  67
Potato  241
Prawns  309–10
Preserving specimens
  methods of  20–9
  materials for  33–4
Propagation *see* Reproduction
Pupae  76–7

Regelation  147
Reproduction  48–50
  asexual  213
  by seeds  214–15, 216, 222–3
  by spores  215–6, 222
  of animals  48–9
    methods  176–80
  of crustaceans  307
  of harvestmen  303
  of plants  49, 59–60, 212–16
  of spiders  300
  vegetative  212–16
Reptiles  199–209
  evolution of  200–02
  respiration  157
Reservoirs  141–2
Respiration  88, 157–8
  of amphibians  157, 187
  of crustaceans  306, 307
  of harvestmen  302
  of plants  229

Respiration contd.
  of reptiles  199, 202
  of spiders  297
Rhizoid  54
Rhizomes  237
Rivers, Erosion by  282–3
Rock, Erosion of  279–84
Rolling  99–100
Roots  237–40
  adventitious  215, 240
  aerial  239
  fibrous  239
  purpose of  238
  tap  239

Salt, Common  150–4
  in the body  152
  in water  151
  sources of  151, 152–3
  uses of  153
Samara  66
Saprohytes  54, 275
Satellites  261, 262, 267, 268
  man-made  267–8
Science table  10–11, 34–7
Sea, Erosion by  283–4
Seaweed  55
Seeds  49, 58, 60
  in fruits  63–7
  reproduction by  214–15
Sepals  62
Shrimps  307
  freshwater  311–12
Shrubs  53, 55
  flowers on  222–7
  leaves on  229–32
  reproduction  222
  stems  236–7
  and buds  171–3
Siliqua  66
Simple plants  53–4, 55–6
  in fresh water  292–3
  reproduction of  212, 213
Sirocco  110
Sleet  112, 149
Sliding  97–8
Snakes  205
Snow  111, 146–7
Solids  39
  in solution  133
  mixtures containing  126–7, 128–9, 132
Solutions  127, 132–4
  saturated  133, 134
Spiders  297–302
  in captivity  304
Spiracles  157
Spores  49, 58, 212, 222
  reproduction by  213
Springs (water)  140
Stamens  62

Stars 260–3
Steam 41
Stems 236–7
  and buds 171–3
  modified 237
  purposes of 236
  underground 237
Stick insects 73, 78–9
Stickleback 181
Stipules 232
Sun 263
  and weather 109–10
  force of gravity of 118

Tadpoles 188
  in captivity 194–6
Tendrils 236–7, 240–1
Tides 118
Toads *see* Amphibians
Tornado 110
Tortoises and turtles 203–4
  as pets 207–9
Trees 53, 55
  flowers on 222–7
  leaves on 229–32
  reproduction 222
  stems 236–7
    and buds 171–3

Typhoon 109

Vapour 41

Water
  and weather 110–12, 117, 138,
    145–9
  as food 137–43, 146
  cycle 138, 151
  fleas 312
  ground 139–41
  in solutions 133
  moving 93–4, 117, 282–4
  purification of 143
  surface 141–3
  table 140
  vapour 110–11, 138, 146
Weather 109–12
  and gravity 117
Weathering 279–84
Weight 120–1
Wells 140–41
Wheels 104–6
Wind 93, 109–10, 117, 281
Woodlice 306, 312–13

Yeast 215